Handbook of
Regression Analysis

Handbook of
Regression Analysis

Samprit Chatterjee
New York University

Jeffrey S. Simonoff
New York University

A JOHN WILEY & SONS, INC., PUBLICATION

For general information on our other products and services or for technical support, please contact our Customer Care Department within the United States at (800) 762-2974, outside the United States at (317) 572-3993 or fax (317) 572-4002.

Wiley also publishes its books in a variety of electronic formats. Some content that appears in print may not be available in electronic formats. For more information about Wiley products, visit our web site at www.wiley.com.

Library of Congress Cataloging-in-Publication Data is available.

ISBN: 978-0-470-88716-5

10 9 8 7 6 5 4 3 2 1

Dedicated to everyone who labors in the field of statistics, whether they are students, teachers, researchers, or data analysts.

CONTENTS

PART II
ADDRESSING VIOLATIONS OF ASSUMPTIONS

PART III
CATEGORICAL PREDICTORS

PREFACE

How to Use This Book

This book is designed to be a practical guide to regression modeling. There is little theory here, and methodology appears in the service of the ultimate goal of analyzing real data using appropriate regression tools. As such, the target audience of the book includes anyone who is faced with regression data [that is, data where there is a response variable that is being modeled as a function of other variable(s)], and whose goal is to learn as much as possible from that data.

The book can be used as a text for an applied regression course (indeed, much of it is based on handouts that have been given to students in such a course), but that is not its primary purpose; rather, it is aimed much more broadly as a source of practical advice on how to address the problems that come up when dealing with regression data. While a text is usually organized in a way that makes the chapters interdependent, successively building on each other, that is not the case here. Indeed, we encourage readers to dip into different chapters for practical advice on specific topics as needed. The pace of the book is faster than might typically be the case for a text. The coverage, while at an applied level, does not shy away from sophisticated concepts. It is distinct from, for example, Chatterjee and Hadi (2012), while also having less theoretical focus than texts such as Greene (2011), Montgomery et al. (2012), or Sen and Srivastava (1990).

This, however, is not a cookbook that presents a mechanical approach to doing regression analysis. Data analysis is perhaps an art, and certainly a craft; we believe that the goal of any data analysis book should be to help analysts develop the skills and experience necessary to adjust to the inevitable twists and turns that come up when analyzing real data.

We assume that the reader possesses a nodding acquaintance with regression analysis. The reader should be familiar with the basic terminology and should have been exposed to basic regression techniques and concepts, at least at the level of simple (one-predictor) linear regression. We also assume that the user has access to a computer with an adequate regression package. The material presented here is not tied to any particular software. Almost all of the analyses described here can be performed by most standard packages, although the ease of doing this could vary. All of the analyses presented here were done using the free package R (R Development Core Team, 2011), which is available for many different operating system platforms (see http://www.R-project.org/ for more information). Code for

the output and figures in the book can be found at its associated web site at
`http://people.stern.nyu.edu/jsimonof/RegressionHandbook/`.

Each chapter of the book is laid out in a similar way, with most having at
least four sections of specific types. First is an introduction, where the general
issues that will be discussed in that chapter are presented. A section on concepts and background material follows, where a discussion of the relationship
of the chapter's material to the broader study of regression data is the focus.
This section also provides any theoretical background for the material that is
necessary. Sections on methodology follow, where the specific tools used in
the chapter are discussed. This is where relevant algorithmic details are likely
to appear. Finally, each chapter includes at least one analysis of real data using the methods discussed in the chapter (as well as appropriate material from
earlier chapters), including both methodological and graphical analyses.

The book begins with discussion of the multiple regression model. Many
regression textbooks start with discussion of simple regression before moving
on to multiple regression. This is quite reasonable from a pedagogical point
of view, since simple regression has the great advantage of being easy to understand graphically, but from a practical point of view simple regression is
rarely the primary tool in analysis of real data. For that reason, we start with
multiple regression, and note the simplifications that come from the special
case of a single predictor. Chapter 1 describes the basics of the multiple regression model, including the assumptions being made, and both estimation
and inference tools, while also giving an introduction to the use of residual
plots to check assumptions.

Since it is unlikely that the first model examined will ultimately be the
final preferred model, Chapter 2 focuses on the very important areas of model
building and model selection. This includes addressing the issue of collinearity, as well as the use of both hypothesis tests and information measures to
help choose among candidate models.

Chapters 3 through 5 study common violations of regression assumptions, and methods available to address those model violations. Chapter 3
focuses on unusual observations (outliers and leverage points), while Chapter
4 describes how transformations (especially the log transformation) can often
address both nonlinearity and nonconstant variance violations. Chapter 5 is
an introduction to time series regression, and the problems caused by autocorrelation. Time series analysis is a vast area of statistical methodology, so
our goal in this chapter is only to provide a good practical introduction to
that area in the context of regression analysis.

Chapters 6 and 7 focus on the situation where there are categorical variables among the predictors. Chapter 6 treats analysis of variance (ANOVA)
models, which include only categorical predictors, while Chapter 7 looks at
analysis of covariance (ANCOVA) models, which include both numerical and
categorical predictors. The examination of interaction effects is a fundamental
aspect of these models, as are questions related to simultaneous comparison of
many groups to each other. Data of this type often exhibit nonconstant vari-

ance related to the different subgroups in the population, and the appropriate tool to address this issue, weighted least squares, is also a focus here.

Chapters 8 though 10 examine the situation where the nature of the response variable is such that Gaussian-based least squares regression is no longer appropriate. Chapter 8 focuses on logistic regression, designed for binary response data and based on the binomial random variable. While there are many parallels between logistic regression analysis and least squares regression analysis, there are also issues that come up in logistic regression that require special care. Chapter 9 uses the multinomial random variable to generalize the models of Chapter 8 to allow for multiple categories in the response variable, outlining models designed for response variables that either do or do not have ordered categories. Chapter 10 focuses on response data in the form of counts, where distributions like the Poisson and negative binomial play a central role. The connection between all these models through the generalized linear model framework is also exploited in this chapter.

The final chapter focuses on situations where linearity does not hold, and a nonlinear relationship is necessary. Although these models are based on least squares, from both an algorithmic and inferential point of view there are strong connections with the models of Chapters 8 through 10, which we highlight.

This Handbook can be used in several different ways. First, a reader may use the book to find information on a specific topic. An analyst might want additional information on, for example, logistic regression or autocorrelation. The chapters on these (and other) topics provide the reader with this subject matter information. As noted above, the chapters also include at least one analysis of a data set, a clarification of computer output, and reference to sources where additional material can be found. The chapters in the book are to a large extent self-contained and can be consulted independently of other chapters.

The book can also be used as a template for what we view as a reasonable approach to data analysis in general. This is based on the cyclical paradigm of model formulation, model fitting, model evaluation, and model updating leading back to model (re)formulation. Statistical significance of test statistics does not necessarily mean that an adequate model has been obtained. Further analysis needs to be performed before the fitted model can be regarded as an acceptable description of the data, and this book concentrates on this important aspect of regression methodology. Detection of deficiencies of fit is based on both testing and graphical methods, and both approaches are highlighted here.

This preface is intended to indicate ways in which the Handbook can be used. Our hope is that it will be a useful guide for data analysts, and will help contribute to effective analyses. We would like to thank our students and colleagues for their encouragement and support. We hope we have provided them with a book of which they would approve. We would like to thank Steve Quigley, Jackie Palmieri, and Amy Hendrickson for their help in bringing this manuscript to print. We would also like to thank our families for their love and support.

SAMPRIT CHATTERJEE
Brooksville, Maine

JEFFREY S. SIMONOFF
New York, New York

August, 2012

PART ONE

The Multiple Linear Regression Model

Multiple Linear Regression

1.1 Introduction

This is a book about regression modeling, but when we refer to regression models, what do we mean? The regression framework can be characterized in the following way:

1. We have one particular variable that we are interested in understanding or modeling, such as sales of a particular product, sale price of a home, or voting preference of a particular voter. This variable is called the **target**, **response**, or **dependent** variable, and is usually represented by y.

2. We have a set of p other variables that we think might be useful in predicting or modeling the target variable (the price of the product, the competitor's price, and so on; or the lot size, number of bedrooms, number of bathrooms of the home, and so on; or the gender, age, income, party membership of the voter, and so on). These are called the **predicting**, or **independent** variables, and are usually represented by x_1, x_2, etc.

Typically, a regression analysis is used for one (or more) of three purposes:

1. modeling the relationship between **x** and y;
2. prediction of the target variable (forecasting);
3. and testing of hypotheses.

In this chapter we introduce the basic multiple linear regression model, and discuss how this model can be used for these three purposes. Specifically, we discuss the interpretations of the estimates of different regression parameters, the assumptions underlying the model, measures of the strength of the relationship between the target and predictor variables, the construction of tests of hypotheses and intervals related to regression parameters, and the checking of assumptions using diagnostic plots.

1.2   Concepts and Background Material

1.2.1 THE LINEAR REGRESSION MODEL

The data consist of n sets of observations $\{x_{1i}, x_{2i}, \ldots, x_{pi}, y_i\}$, which represent a random sample from a larger population. It is assumed that these observations satisfy a linear relationship,

$$y_i = \beta_0 + \beta_1 x_{1i} + \cdots + \beta_p x_{pi} + \varepsilon_i, \tag{1.1}$$

where the β coefficients are unknown parameters, and the ε_i are random error terms. By a *linear* model, it is meant that the model is linear in the *parameters*; a quadratic model,

$$y_i = \beta_0 + \beta_1 x_i + \beta_2 x_i^2 + \varepsilon_i,$$

paradoxically enough, is a linear model, since x and x^2 are just versions of x_1 and x_2.

It is important to recognize that this, or any statistical model, is not viewed as a *true* representation of reality; rather, the goal is that the model be a *useful* representation of reality. A model can be used to explore the relationships between variables and make accurate forecasts based on those relationships even if it is not the "truth." Further, any statistical model is only temporary, representing a provisional version of views about the random process being studied. Models can, and should, change, based on analysis using the current model, selection among several candidate models, the acquisition

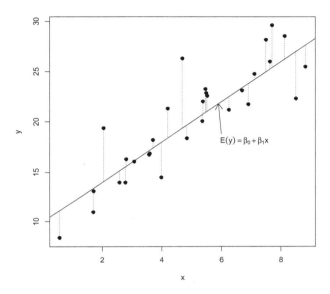

FIGURE 1.1 The simple linear regression model. The solid line corresponds to the true regression line, and the dotted lines correspond to the random errors ε_i.

of new data, and so on. Further, it is often the case that there are several different models that are reasonable representations of reality. Having said this, we will sometimes refer to the "true" model, but this should be understood as referring to the underlying form of the currently hypothesized representation of the regression relationship.

The special case of (1.1) with $p = 1$ corresponds to the **simple regression model**, and is consistent with the representation in Figure 1.1. The solid line is the true regression line, the expected value of y given the value of x. The dotted lines are the random errors ε_i that account for the lack of a perfect association between the predictor and the target variables.

1.2.2 ESTIMATION USING LEAST SQUARES

The true regression function represents the expected relationship between the target and the predictor variables, which is unknown. A primary goal of a regression analysis is to estimate this relationship, or equivalently, to estimate the unknown parameters β. This requires a data-based rule, or criterion, that will give a reasonable estimate. The standard approach is **least squares**

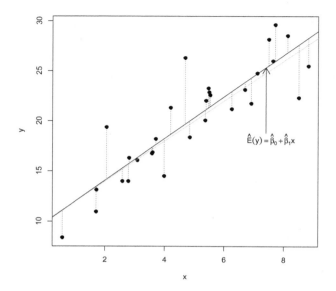

FIGURE 1.2 Least squares estimation for the simple linear regression model, using the same data as in Figure 1.1. The gray line corresponds to the true regression line, the solid black line corresponds to the fitted least squares line (designed to estimate the gray line), and the lengths of the dotted lines correspond to the residuals. The sum of squared values of the lengths of the dotted lines is minimized by the solid black line.

regression, where the estimates are chosen to minimize

$$\sum_{i=1}^{n}[y_i - (\beta_0 + \beta_1 x_{1i} + \cdots + \beta_p x_{pi})]^2. \tag{1.2}$$

Figure 1.2 gives a graphical representation of least squares that is based on Figure 1.1. Now the true regression line is represented by the gray line, and the solid black line is the estimated regression line, designed to estimate the (unknown) gray line as closely as possible. For any choice of estimated parameters $\hat{\beta}$, the estimated expected response value given the observed predictor values equals

$$\hat{y}_i = \hat{\beta}_0 + \hat{\beta}_1 x_{1i} + \cdots + \hat{\beta}_p x_{pi},$$

and is called the **fitted value.** The difference between the observed value y_i and the fitted value $\hat{y}_i$ is called the **residual,** the set of which are represented by the lengths of the dotted lines in Figure 1.2. The least squares regression line minimizes the sum of squares of the lengths of the dotted lines; that is, the ordinary least squares (OLS) estimates minimize the sum of squares of the residuals.

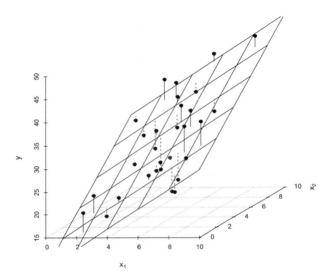

FIGURE 1.3 Least squares estimation for the multiple linear regression model with two predictors. The plane corresponds to the fitted least squares relationship, and the lengths of the vertical lines correspond to the residuals. The sum of squared values of the lengths of the vertical lines is minimized by the plane.

In higher dimensions ($p > 1$) the true and estimated regression relationships correspond to planes ($p = 2$) or hyperplanes ($p \geq 3$), but otherwise the principles are the same. Figure 1.3 illustrates the case with two predictors. The length of each vertical line corresponds to a residual (solid lines refer to positive residuals while dashed lines refer to negative residuals), and the (least squares) plane that goes through the observations is chosen to minimize the sum of squares of the residuals.

The linear regression model can be written compactly using matrix notation. Define the following matrix and vectors as follows:

$$X = \begin{pmatrix} 1 & x_{11} & \cdots & x_{p1} \\ \vdots & \vdots & & \vdots \\ 1 & x_{1n} & \cdots & x_{pn} \end{pmatrix} \quad \mathbf{y} = \begin{pmatrix} y_1 \\ \vdots \\ y_n \end{pmatrix} \quad \boldsymbol{\beta} = \begin{pmatrix} \beta_0 \\ \beta_1 \\ \vdots \\ \beta_p \end{pmatrix} \quad \boldsymbol{\varepsilon} = \begin{pmatrix} \varepsilon_1 \\ \vdots \\ \varepsilon_n \end{pmatrix}.$$

The regression model (1.1) is then

$$\mathbf{y} = X\boldsymbol{\beta} + \boldsymbol{\varepsilon}.$$

The normal equations [which determine the minimizer of (1.2)] can be shown (using multivariate calculus) to be

$$(X'X)\hat{\boldsymbol{\beta}} = X'\mathbf{y},$$

which implies that the least squares estimates satisfy

$$\hat{\boldsymbol{\beta}} = (X'X)^{-1}X'\mathbf{y}.$$

The fitted values are then

$$\hat{\mathbf{y}} = X\hat{\boldsymbol{\beta}} = X(X'X)^{-1}X'\mathbf{y} \equiv H\mathbf{y}, \tag{1.3}$$

where $H = X(X'X)^{-1}X'$ is the so-called "hat" matrix (since it takes $\mathbf{y}$ to $\hat{\mathbf{y}}$). The residuals $\mathbf{e} = \mathbf{y} - \hat{\mathbf{y}}$ thus satisfy

$$\mathbf{e} = \mathbf{y} - \hat{\mathbf{y}} = \mathbf{y} - X(X'X)^{-1}X'\mathbf{y} = (I - X(X'X)^{-1}X')\mathbf{y}, \tag{1.4}$$

or

$$\mathbf{e} = (I - H)\mathbf{y}.$$

1.2.3 ASSUMPTIONS

The least squares criterion will not necessarily yield sensible results unless certain assumptions hold. One is given in (1.1) — the linear model should be appropriate. In addition, the following assumptions are needed to justify using least squares regression.

1. The expected value of the errors is zero ($E(\varepsilon_i) = 0$ for all i). That is, it cannot be true that for certain observations the model is systematically too low, while for others it is systematically too high. A violation of this assumption will lead to difficulties in estimating β_0. More importantly, this reflects that the model does not include a necessary systematic component, which has instead been absorbed into the error terms.

2. The variance of the errors is constant ($V(\varepsilon_i) = \sigma^2$ for all i). That is, it cannot be true that the strength of the model is more for some parts of the population (smaller σ) and less for other parts (larger σ). This assumption of constant variance is called **homoscedasticity**, and its violation (nonconstant variance) is called **heteroscedasticity**. A violation of this assumption means that the least squares estimates are not as efficient as they could be in estimating the true parameters, and better estimates are available. More importantly, it also results in poorly calibrated confidence and (especially) prediction intervals.

3. The errors are uncorrelated with each other. That is, it cannot be true that knowing that the model underpredicts y (for example) for one particular observation says anything at all about what it does for any other

observation. This violation most often occurs in data that are ordered in time (time series data), where errors that are near each other in time are often similar to each other (such time-related correlation is called **autocorrelation**). Violation of this assumption can lead to very misleading assessments of the strength of the regression.

4. The errors are normally distributed. This is needed if we want to construct any confidence or prediction intervals, or hypothesis tests, which we usually do. If this assumption is violated, hypothesis tests and confidence and prediction intervals can be very misleading.

Since violation of these assumptions can potentially lead to completely misleading results, a fundamental part of any regression analysis is to check them using various plots, tests, and diagnostics.

1.3 Methodology

1.3.1 INTERPRETING REGRESSION COEFFICIENTS

The least squares regression coefficients have very specific meanings. They are often misinterpreted, so it is important to be clear on what they mean (and do not mean). Consider first the intercept, $\hat{\beta}_0$.

$\hat{\beta}_0$: The estimated expected value of the target variable when the predictors all equal zero.

Note that this might not have any physical interpretation, since a zero value for the predictor(s) might be impossible, or might never come close to occurring in the observed data. In that situation, it is pointless to try to interpret this value. If all of the predictors are centered to have mean zero, then $\hat{\beta}_0$ necessarily equals $\overline{Y}$, the sample mean of the target values. Note that if there is any particular value for each predictor that is meaningful in some sense, if each variable is centered around its particular value, then the intercept is an estimate of $E(y)$ when the predictors all have those meaningful values.

The estimated coefficient for the jth predictor ($j = 1, \ldots, p$) is interpreted in the following way.

$\hat{\beta}_j$: The estimated expected change in the target variable associated with a one unit change in the jth predicting variable, holding all else in the model fixed.

There are several noteworthy aspects to this interpretation. First, note the word *associated* — we cannot say that a change in the target variable is *caused* by a change in the predictor, only that they are associated with each other. That is, correlation does not imply causation.

Another key point is the phrase "holding all else in the model fixed," the implications of which are often ignored. Consider the following hypothetical example. A random sample of college students at a particular university is taken in order to understand the relationship between college grade point average (GPA) and other variables. A model is built with college GPA as a function of high school GPA and the standardized Scholastic Aptitude Test (SAT), with resultant least squares fit

$$\texttt{College GPA} = 1.3 + .7 \times \texttt{High School GPA} - .0001 \times \texttt{SAT}.$$

It is tempting to say (and many people would say) that the coefficient for SAT score has the "wrong sign," because it says that higher values of SAT are associated with lower values of college GPA. This is *not* correct. The problem is that it is likely in this context that what an analyst would find intuitive is the *marginal* relationship between college GPA and SAT score alone (ignoring all else), one that we would indeed expect to be a direct (positive) one. The regression coefficient does not say anything about that marginal relationship. Rather, it refers to the conditional (sometimes called partial) relationship that takes the high school GPA as fixed, which is apparently that higher values of SAT are associated with lower values of college GPA, holding high school GPA fixed. High school GPA and SAT are no doubt related to each other, and it is quite likely that this relationship between the predictors would complicate any understanding of, or intuition about, the conditional relationship between college GPA and SAT score. Multiple regression coefficients should not be interpreted marginally; if you really are interested in the relationship between the target and a single predictor alone, you should simply do a regression of the target on only that variable. This does not mean that multiple regression coefficients are uninterpretable, only that care is necessary when interpreting them.

Another common use of multiple regression that depends on this conditional interpretation of the coefficients is to explicitly include "control" variables in a model in order to try to account for their effect statistically. This is particularly important in observational data (data that are not the result of a designed experiment), since in that case the effects of other variables cannot be ignored as a result of random assignment in the experiment. For observational data it is not possible to physically intervene in the experiment to "hold other variables fixed," but the multiple regression framework effectively allows this to be done statistically.

1.3.2 MEASURING THE STRENGTH OF THE REGRESSION RELATIONSHIP

The least squares estimates possess an important property:

$$\sum_{i=1}^{n}(y_i - \overline{Y})^2 = \sum_{i=1}^{n}(y_i - \hat{y}_i)^2 + \sum_{i=1}^{n}(\hat{y}_i - \overline{Y})^2.$$

This formula says that the variability in the target variable (the left side of the equation, termed the corrected total sum of squares) can be split into two mutually exclusive parts — the variability left over after doing the regression (the first term on the right side, the residual sum of squares), and the variability accounted for by doing the regression (the second term, the regression sum of squares). This immediately suggests the usefulness of R^2 as a measure of the strength of the regression relationship, where

$$R^2 = \frac{\sum_i(\hat{y}_i - \overline{Y})^2}{\sum_i(y_i - \overline{Y})^2} \equiv \frac{\text{Regression SS}}{\text{Corrected total SS}} = 1 - \frac{\text{Residual SS}}{\text{Corrected total SS}}.$$

The R^2 value (also called the **coefficient of determination**) estimates the population proportion of variability in y accounted for by the best linear combination of the predictors. Values closer to 1 indicate a good deal of predictive power of the predictors for the target variable, while values closer to 0 indicate little predictive power. An equivalent representation of R^2 is

$$R^2 = \text{corr}(y_i, \hat{y}_i)^2,$$

where

$$\text{corr}(y_i, \hat{y}_i) = \frac{\sum_i(y_i - \overline{Y})(\hat{y}_i - \overline{\hat{Y}})}{\sqrt{\sum_i(y_i - \overline{Y})^2 \sum_i(\hat{y}_i - \overline{\hat{Y}})^2}}$$

is the sample correlation coefficient between **y** and **ŷ** (this correlation is called the multiple correlation coefficient). That is, R^2 is a direct measure of how similar the observed and fitted target values are.

It can be shown that R^2 is biased upwards as an estimate of the population proportion of variability accounted for by the regression. The **adjusted** R^2 corrects this bias, and equals

$$R_a^2 = R^2 - \frac{p}{n-p-1}\left(1 - R^2\right). \tag{1.5}$$

It is apparent from (1.5) that unless p is large relative to $n - p - 1$ (that is, unless the number of predictors is large relative to the sample size), R^2 and R_a^2 will be close to each other, and the choice of which to use is a minor concern. What is perhaps more interesting is the nature of R_a^2 as providing an explicit tradeoff between the strength of the fit (the first term, with larger R^2 corresponding to stronger fit and larger R_a^2) and the complexity of the model (the second term, with larger p corresponding to more complexity and smaller R_a^2). This tradeoff of fidelity to the data versus simplicity will be important in the discussion of model selection in Section 2.3.1.

The only parameter left unaccounted for in the estimation scheme is the variance of the errors σ^2. An unbiased estimate is provided by the **residual mean square**,

$$\hat{\sigma}^2 = \frac{\sum_{i=1}^{n}(y_i - \hat{y}_i)^2}{n-p-1}. \tag{1.6}$$

This estimate has a direct, but often underappreciated, use in assessing the practical importance of the model. Does knowing $x_1, \ldots, x_p$ really say anything of value about y? This isn't a question that can be answered completely statistically; it requires knowledge and understanding of the data (that is, it requires context). Recall that the model assumes that the errors are normally distributed with standard deviation σ. This means that, roughly speaking, 95% of the time an observed y value falls within $\pm 2\sigma$ of the expected response

$$E(y) = \beta_0 + \beta_1 x_1 + \cdots + \beta_p x_p.$$

$E(y)$ can be estimated for any given set of $\mathbf{x}$ values using

$$\hat{y} = \hat{\beta}_0 + \hat{\beta}_1 x_1 + \cdots + \hat{\beta}_p x_p,$$

while the square root of the residual mean square (1.6), termed the **standard error of the estimate**, provides an estimate of σ that can be used in constructing this rough prediction interval $\pm 2\hat{\sigma}$.

1.3.3 HYPOTHESIS TESTS AND CONFIDENCE INTERVALS FOR β

There are two types of hypothesis tests related to the regression coefficients of immediate interest.

1. Do *any* of the predictors provide predictive power for the target variable? This is a test of the overall significance of the regression,

$$H_0 : \beta_1 = \cdots = \beta_p = 0$$

versus
$$H_a : \text{some } \beta_j \neq 0, \qquad j = 1, \ldots, p.$$

The test of these hypotheses is the F-**test**,

$$F = \frac{\text{Regression MS}}{\text{Residual MS}} \equiv \frac{\text{Regression SS}/p}{\text{Residual SS}/(n-p-1)}.$$

This is referenced against a null F-distribution on $(p, n-p-1)$ degrees of freedom.

2. Given the other variables in the model, does a particular predictor provide additional predictive power? This corresponds to a test of the significance of an individual coefficient,

$$H_0 : \beta_j = 0, \qquad j = 1, \ldots, p$$

versus
$$H_a : \beta_j \neq 0.$$

This is tested using a t-**test**,

$$t_j = \frac{\hat{\beta}_j}{\widehat{s.e.}(\hat{\beta}_j)},$$

which is compared to a t-distribution on $n - p - 1$ degrees of freedom. Other values of β_j can be specified in the null hypothesis (say β_{j0}), with the t-statistic becoming

$$t_j = \frac{\hat{\beta}_j - \beta_{j0}}{\widehat{s.e.}(\hat{\beta}_j)}. \tag{1.7}$$

The values of $\widehat{s.e.}(\hat{\beta}_j)$ are obtained as the square roots of the diagonal elements of $\hat{V}(\hat{\beta}) = (X'X)^{-1}\hat{\sigma}^2$, where $\hat{\sigma}^2$ is the residual mean square (1.6). Note that for simple regression ($p = 1$) the hypotheses corresponding to the overall significance of the model and the significance of the predictor are identical,

$$H_0 : \beta_1 = 0$$

versus

$$H_a : \beta_1 \neq 0.$$

Given the equivalence of the sets of hypotheses, it is not surprising that the associated tests are also equivalent; in fact, $F = t_1^2$, and the associated tail probabilities of the two tests are identical.

A t-test for the intercept also can be constructed as in (1.7), although this does not refer to a hypothesis about a predictor, but rather about whether the expected target is equal to a specified value β_{00} if all of the predictors equal zero. As was noted in Section 1.3.1, this is often not physically meaningful (and therefore of little interest), because the condition that all predictors equal zero cannot occur, or does not come close to occurring in the observed data.

As is always the case, a confidence interval provides an alternative way of summarizing the degree of precision in the estimate of a regression parameter. That is, a $100 \times (1 - \alpha)\%$ confidence interval for β_j has the form

$$\hat{\beta}_j \pm t_{\alpha/2}^{n-p-1}\widehat{s.e.}(\hat{\beta}_j),$$

where $t_{\alpha/2}^{n-p-1}$ is the appropriate critical value at two-sided level α for a t-distribution on $n - p - 1$ degrees of freedom.

1.3.4 FITTED VALUES AND PREDICTIONS

The rough prediction interval $\hat{y} \pm 2\hat{\sigma}$ discussed in Section 1.3.2 is an approximate 95% interval because it ignores the variability caused by the need to estimate σ, and uses only an approximate normal-based critical value. A more accurate assessment of this is provided by a **prediction interval** given a particular value of $\mathbf{x}$. This interval provides guidance as to how precise $\hat{y}_0$ is as a prediction of y for some particular specified value $\mathbf{x}_0$, where $\hat{y}_0$ is determined by substituting the values $\mathbf{x}_0$ into the estimated regression equation; its width depends on both $\hat{\sigma}$ and the position of $\mathbf{x}_0$ relative to the centroid of the predictors (the point located at the means of all predictors), since values farther from the centroid are harder to predict as precisely. Specifically, for a simple regression, the estimated standard error of a predicted value based on a value x_0 of the predicting variable is

$$\widehat{s.e.}(\hat{y}_0^P) = \hat{\sigma}\sqrt{1 + \frac{1}{n} + \frac{(x_0 - \overline{X})^2}{\sum(x_i - \overline{X})^2}}.$$

More generally, the variance of a predicted value is

$$\hat{V}(\hat{y}_0^P) = [1 + \mathbf{x}_0'(X'X)^{-1}\mathbf{x}_0]\hat{\sigma}^2. \tag{1.8}$$

Here $\mathbf{x}_0$ is taken to include a 1 in the first entry (corresponding to the intercept in the regression model). The prediction interval is then

$$\hat{y}_0 \pm t_{\alpha/2}^{n-p-1}\widehat{s.e.}(\hat{y}_0^P),$$

where $\widehat{s.e.}(\hat{y}_0^P) = \sqrt{\hat{V}(\hat{y}_0^P)}$.

This prediction interval should not be confused with a **confidence interval for a fitted value**. The prediction interval is used to provide an interval estimate for a *prediction of y for one member of the population* with a particular value of $\mathbf{x}_0$; the confidence interval is used to provide an interval estimate for the *true expected value of y for all members of the population* with a particular value of $\mathbf{x}_0$. The corresponding standard error, termed the standard error for a fitted value, is the square root of

$$\hat{V}(\hat{y}_0^F) = \mathbf{x}_0'(X'X)^{-1}\mathbf{x}_0\hat{\sigma}^2, \tag{1.9}$$

with corresponding confidence interval

$$\hat{y}_0 \pm t_{\alpha/2}^{n-p-1}\widehat{s.e.}(\hat{y}_0^F).$$

A comparison of the two estimated variances (1.8) and (1.9) shows that the variance of the predicted value has an extra σ^2 term, which corresponds to the inherent variability in the population. Thus, the confidence interval for a fitted value will always be narrower than the prediction interval, and is often much narrower (especially for large samples), since increasing the sample size will always improve estimation of the expected response value, but cannot lessen the inherent variability in the population associated with the prediction of the target for a single observation.

1.3.5 CHECKING ASSUMPTIONS USING RESIDUAL PLOTS

As was noted earlier, all of these tests, intervals, predictions, and so on, are based on the belief that the assumptions of the regression model hold. Thus, it is crucially important that these assumptions be checked. Remarkably enough, a few very simple plots can provide much of the evidence needed to check the assumptions.

1. A plot of the residuals versus the fitted values. This plot should have no pattern to it; that is, no structure should be apparent. Certain kinds of structure indicate potential problems:

   (a) A point (or a few points) isolated at the top or bottom, or left or right. In addition, often the rest of the points have a noticeable "tilt" to them. These isolated points are unusual points, and can have a strong effect on the regression. They need to be examined carefully, and possibly removed from the data set.

   (b) An impression of different heights of the point cloud as the plot is examined from left to right. This indicates potential heteroscedasticity (nonconstant variance).

2. Plots of the residuals versus each of the predictors. Again, a plot with no apparent structure is desired.

3. If the data set has a time structure to it, residuals should be plotted versus time. Again, there should be no apparent pattern. If there is a cyclical structure, this indicates that the errors are not uncorrelated, as they are supposed to be (that is, there is potentially autocorrelation in the errors).

4. A normal plot of the residuals. This plot assesses the apparent normality of the residuals, by plotting the observed ordered residuals on one axis and the expected positions (under normality) of those ordered residuals on the other. The plot should look like a straight line (roughly). Isolated points once again represent unusual observations, while a curved line indicates that the errors are probably not normally distributed, and tests and intervals might not be trustworthy.

Note that all of these plots should be routinely examined in any regression analysis, although in order to save space not all will necessarily be presented in all of the analyses in the book.

An implicit assumption in any model that is being used for prediction is that the future "looks like" the past; that is, it is not sufficient that these assumptions appear to hold for the available data, as they also must continue to hold for new data on which the estimated model is applied. Indeed, the assumption is stronger than that, since it must be the case that the future is exactly the same as the past, in the sense that all of the properties of the model, including the precise values of all of the regression parameters, are the same. This is unlikely to be exactly true, so a more realistic point of view is that the future should be similar enough to the past so that predictions based on the past are useful. A related point is that predictions should not be based on extrapolation, where the predictor values are far from the values used to build the model. Similarly, if the observations form a time series, predictions far into the future are unlikely to be very useful.

In general, the more complex a model is, the less likely it is that all of its characteristics will remain stable going forward, which implies that a reasonable goal is to try to find a model that is as simple as it can be while still accounting for the important effects in the data. This leads to questions of **model building**, which is the subject of the next chapter.

1.4   Example — Estimating Home Prices

Determining the appropriate sale price for a home is clearly of great interest to both buyers and sellers. While this can be done in principle by examining the prices at which other similar homes have recently sold, the well-known existence of strong effects related to location means that there are likely to be relatively few homes with the same important characteristics to make the comparison. A solution to this problem is the use of hedonic regression models, where the sale prices of a set of homes in a particular area are regressed on important characteristics of the home such as the number of bedrooms, the living area, the lot size, and so on. Academic research on this topic is plentiful, going back to at least Wabe (1971).

This analysis is based on a sample from public data on sales of one-family homes in the Levittown, NY area from June 2010 through May 2011. Levittown is famous as the first planned suburban community built using mass production methods, being aimed at former members of the military after World War II. Most of the homes in this community were built in the late 1940s to early 1950s, without basements and designed to make expansion on the second floor relatively easy.

For each of the 85 houses in the sample, the number of bedrooms, number of bathrooms, living area (in square feet), lot size (in square feet), the year the house was built, and the property taxes are used as potential predictors of the sale price. In any analysis the first step is to look at the data, and Figure 1.4 gives scatter plots of sale price versus each predictor. It is apparent that there is a positive association between sale price and each variable, other than number of bedrooms and lot size. We also note that there are two houses with unusually large living areas for this sample, two with unusually large property taxes (these are not the same two houses), and three that were built 6 or 7 years later than all of the other houses in the sample.

The output below summarizes the results of a multiple regression fit.

```
Coefficients:
              Estimate Std. Error t value Pr(>|t|)
(Intercept) -7.149e+06  3.820e+06  -1.871 0.065043 .
Bedrooms    -1.229e+04  9.347e+03  -1.315 0.192361
Bathrooms    5.170e+04  1.309e+04   3.948 0.000171 ***
Living.area  6.590e+01  1.598e+01   4.124 9.22e-05 ***
Lot.size    -8.971e-01  4.194e+00  -0.214 0.831197
Year.built   3.761e+03  1.963e+03   1.916 0.058981 .
Property.tax 1.476e+00  2.832e+00   0.521 0.603734
---
```

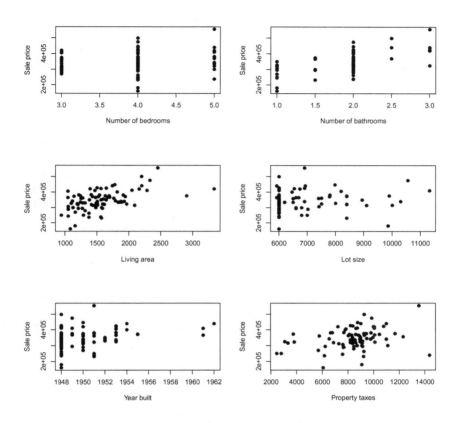

FIGURE 1.4 Scatter plots of sale price versus each predictor for the home price data.

```
Signif. codes:
  0 '***' 0.001 '**' 0.01 '*' 0.05 '.' 0.1 ' ' 1

Residual standard error: 47380 on 78 degrees of freedom
Multiple R-squared: 0.5065,     Adjusted R-squared: 0.4685
F-statistic: 13.34 on 6 and 78 DF,  p-value: 2.416e-10
```

The overall regression is strongly statistically significant, with the tail probability of the F-test roughly 10^{-10}. The predictors account for roughly 50% of the variability in sale prices ($R^2 \approx 0.5$). Two of the predictors (number of bathrooms and living area) are highly statistically significant, with tail probabilities less than .0002, and the coefficient of the year built variable is marginally statistically significant. The coefficients imply that given all else in the model is held fixed, one additional bathroom in a house is associated with an estimated expected price that is $51,700 higher; one additional square foot of living area is associated with an estimated expected price that is $65.90 higher (given the typical value of the living area variable, a more meaningful statement would probably be that an additional 100 square feet of living area is associated with an estimated expected price that is $659 higher); and a house being built one year later is associated with an estimated expected price that is $3761 higher.

This is a situation where the distinction between a confidence interval for a fitted value and a prediction interval (and which is of more interest to a particular person) is clear. Consider a house with 3 bedrooms, 1 bathroom, 1050 square feet of living area, 6000 square foot lot size, built in 1948, with $6306 in property taxes. Substituting those values into the above equation gives an estimated expected sale price of a house with these characteristics equal to $265,360. A buyer or a seller is interested in the sale price of one particular house, so a prediction interval for the sale price would provide a range for what the buyer can expect to pay and the seller expect to get. The standard error of the estimate $\hat{\sigma} = \$47,380$ can be used to construct a rough prediction interval, in that roughly 95% of the time a house with these characteristics can be expected to sell for within $\pm(2)(47380) = \pm\$94,360$ of that estimated sale price, but a more exact interval might be required. On the other hand, a home appraiser or tax assessor is more interested in the typical (average) sale price for all homes of that type in the area, so they can give a justifiable interval estimate giving the precision of the estimate of the true expected value of the house, so a confidence interval for the fitted value is desired.

Exact 95% intervals for a house with these characteristics can be obtained from statistical software, and turn out to be ($167277, $363444) for the prediction interval and ($238482, $292239) for the confidence interval. As expected, the prediction interval is much wider than the confidence interval, since it reflects the inherent variability in sale prices in the population of houses; indeed, it is probably too wide to be of any practical value in this case, but an interval with smaller coverage (that is expected to include the actual price only 50% of the time, say) might be useful (a 50% interval in this case would be ($231974, $298746), so a seller could be told that there is a 50/50 chance that their house will sell for a value in this range).

The validity of all of these results depends on whether the assumptions hold. Figure 1.5 gives a scatter plot of the residuals versus the fitted values and a normal plot of the residuals for this model fit. There is no apparent pattern in the plot of residuals versus fitted values, and the ordered residuals form a roughly straight line in the normal plot, so there are no apparent violations of assumptions here. The plot of residuals versus each of the predictors (Figure 1.6) also does not show any apparent patterns, other than the houses with unusual living area and year being built, respectively. It would be reasonable to omit these observations to see if they have had an effect on the regression, but we will postpone discussion of that to Chapter 3, where diagnostics for unusual observations are discussed in greater detail.

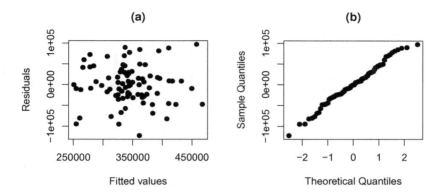

FIGURE 1.5 **Residual plots for the home price data. (a) Plot of residuals versus fitted values. (b) Normal plot of the residuals.**

An obvious consideration at this point is that the models discussed here appear to be overspecified; that is, they include variables that do not apparently add to the predictive power of the model. As was noted earlier, this suggests the consideration of model building, where a more appropriate (simplified) model can be chosen, which will be discussed in the next chapter.

1.5 Summary

In this chapter we have laid out the basic structure of the linear regression model, including the assumptions that justify the use of least squares estimation. The three main goals of regression noted at the beginning of the chapter provide a framework for an organization of the topics covered.

1. Modeling the relationship between **x** and y:

- the least squares estimates $\hat{\beta}$ summarize the expected change in y for a given change in an x, accounting for all of the variables in the model;
- the standard error of the estimate $\hat{\sigma}$ estimates the standard deviation of the errors;
- R^2 and R_a^2 estimate the proportion of variability in y accounted for by **x**;
- and the confidence interval for a fitted value provides a measure of the precision in estimating the expected target for a given set of predictor values.

2. Prediction of the target variable:

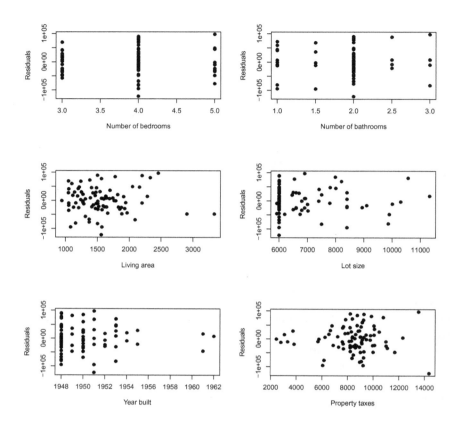

FIGURE 1.6 **Scatter plots of residuals versus each predictor for the home price data.**

- substituting specified values of **x** into the fitted regression model gives an estimate of the value of the target for a new observation;
- the rough prediction interval $\pm 2\hat{\sigma}$ provides a quick measure of the limits of the ability to predict a new observation;
- and the exact prediction interval provides a more precise measure of those limits.

3. Testing of hypotheses:

- the F-test provides a test of the statistical significance of the overall relationship;
- the t-test for each slope coefficient testing whether the true value is zero provides a test of whether the variable provides additional predictive power given the other variables;
- and the t-tests can be generalized to test other hypotheses of interest about the coefficients as well.

Since all of these methods depend on the assumptions holding, a fundamental part of any regression analysis is to check those assumptions. The residual plots discussed in this chapter are a key part of that process, and other diagnostics and tests will be discussed in future chapters that provide additional support for that task.

KEY TERMS

Autocorrelation: Correlation between adjacent observations in a (time) series. In the regression context it is autocorrelation of the errors that is a violation of assumptions.

Coefficient of determination (R^2): The square of the multiple correlation coefficient, estimates the proportion of variability in the target variable that is explained by the fitted least squares model.

Confidence interval for a fitted value: A measure of precision of the estimate of the expected target value for a given **x**.

Dependent variable: Characteristic of each member of the sample that is being modeled. This is also known as the **target** or **response** variable.

Fitted value: The least square estimate of the expected target value for a particular observation obtained from the fitted regression model.

Heteroscedasticity: Unequal variance; this can refer to observed unequal variance of the residuals or theoretical unequal variance of the errors.

Homoscedasticity: Equal variance; this can refer to observed equal variance of the residuals or the assumed equal variance of the errors.

Independent variable(s): Characteristic(s) of each member of the sample that could be used to model the dependent variable. These are also known as the **predicting** variables.

Least squares: A method of estimation that minimizes the sum of squared deviations of the observed target values from their estimated expected values.

Prediction interval: The interval estimate for the value of the target variable for an individual member of the population using the fitted regression model.

Residual: The difference between the observed target value and the corresponding fitted value.

Residual mean square: An unbiased estimate of the variance of the errors. It is obtained by dividing the sum of squares of the residuals by $(n - p - 1)$, where n is the number of observations and p is the number of predicting variables.

Standard error of the estimate ($\hat{\sigma}$): An estimate of σ, the standard deviation of the errors, equaling the square root of the residual mean square.

Model Building

2.1   Introduction

All of the discussion in Chapter 1 is based on the premise that the only model being considered is the one currently being fit. This is not a good data analysis strategy, for several reasons.

1. Including unnecessary predictors in the model (what is sometimes called **overfitting**) complicates descriptions of the process. Using such models tends to lead to poorer predictions because of the additional unnecessary noise. Further, a more complex representation of the true regression relationship is less likely to remain stable enough to be useful for future prediction than is a simpler one.

2. Omitting important effects (**underfitting**) reduces predictive power, biases estimates of effects for included predictors, and results in less understanding of the process being studied.

3. Violations of assumptions should be addressed, so that least squares estimation is justified.

The last of these reasons is the subject of later chapters, while the first two are discussed in this chapter. This operation of choosing among different candidate models so as to avoid overfitting and underfitting is called **model selection**.

First, we discuss the uses of hypothesis testing for model selection. Various hypothesis tests address relevant model selection questions, but there are also reasons why they are not sufficient for these purposes. Part of these difficulties is the effect of correlations among the predictors, and the situation of high correlation among the predictors (**collinearity**) is a particularly challenging one.

A useful way of thinking about the tradeoffs of overfitting versus underfitting noted above is as a contrast between strength of fit and simplicity. The **principle of parsimony** states that a model should be as simple as possible while still accounting for the important relationships in the data. Thus, a sensible way of comparing models is using measures that explicitly reflect this tradeoff; such measures are discussed in Section 2.3.1.

The chapter concludes with a discussion of techniques designed to address the existence of well-defined subgroups in the data. In this situation it is often the case that the effects of a predictor on the target variable is different in the two groups, and ways of building models to handle this are discussed in Section 2.4.

2.2   Concepts and Background Material

2.2.1   USING HYPOTHESIS TESTS TO COMPARE MODELS

Determining whether individual regression coefficients are statistically significant as discussed in Section 1.3.3 is an obvious first step in deciding whether a model is overspecified. A predictor that does not add significantly to model fit should have an estimated slope coefficient that is not significantly different from 0, and is thus identified by a small t-statistic. So, for example, in the analysis of home prices in Section 1.4, the regression output on page 16 suggests removing number of bedrooms, lot size, and property taxes from the model, as all three have insignificant t-values.

It should be noted, however, that t-tests can only assess the contribution of a predictor given all of the others in the model. When predictors are correlated with each other, t-tests can give misleading indications of the importance of a predictor. Consider a two-predictor situation where the predictors are each highly correlated with the target variable, and are also highly correlated with each other. In this situation it is likely that the t-statistic for each predictor will be relatively small. This is not an inappropriate result, since given one predictor the other adds little (being highly correlated with each other, one is redundant in the presence of the other). This means that the t-statistics are not effective in identifying important predictors when the two variables are highly correlated.

The t-tests and F-test of Section 1.3.3 are special cases of a general formulation that is useful for comparing certain classes of models. It might be the case that a simpler version of a candidate model (a *subset* model) might be adequate to fit the data. For example, consider taking a sample of college students and determining their College grade point average (GPA), SAT reading score (Reading) and SAT math score (Math). The *full* regression model to fit to these data is

$$\texttt{GPA}_i = \beta_0 + \beta_1 \texttt{Reading}_i + \beta_2 \texttt{Math}_i + \varepsilon_i.$$

Instead of considering reading and math scores separately, we could consider whether GPA can be predicted by one variable: *total* SAT score, which is the sum of Reading and Math. This *subset* model is

$$\texttt{GPA}_i = \gamma_0 + \gamma_1 (\texttt{Reading} + \texttt{Math})_i + \varepsilon_i,$$

with $\beta_1 = \beta_2 \equiv \gamma_1$. This equality condition is called a **linear restriction**, because it defines a linear condition on the parameters of the regression model (that is, it only involves additions, subtractions, and equalities of coefficients and constants).

The question about whether the total SAT score is sufficient to predict grade point average can be stated using a hypothesis test about this linear restriction. As always, the null hypothesis gets the benefit of the doubt; in this case, that is the *simpler* restricted (subset) model that the sum of Reading and Math is adequate, since it says that only one predictor is needed, rather than two. The alternative hypothesis is the unrestricted full model (with no conditions on β). That is,

$$H_0 : \beta_1 = \beta_2$$

versus

$$H_a : \beta_1 \neq \beta_2.$$

These hypotheses are tested using a **partial F-test**. The F-statistic has the form

$$F = \frac{(\text{Residual SS}_{\text{subset}} - \text{Residual SS}_{\text{full}})/d}{\text{Residual SS}_{\text{full}}/(n - p - 1)}, \qquad (2.1)$$

where n is the sample size, p is the number of predictors in the full model, and d is the difference between the number of parameters in the full model and the number of parameters in the subset model. This statistic is compared to an F distribution on $(d, n - p - 1)$ degrees of freedom. So, for example, for this GPA/SAT example, $p = 2$ and $d = 3 - 2 = 1$, so the observed F-statistic would be compared to an F distribution on $(1, n - 3)$ degrees of freedom. Some statistical packages allow specification of the full and subset models and will calculate the F-test, but others do not, and the statistic has to be calculated manually based on the fits of the two models.

An alternative form for the F-test above might make clearer what is going on here:

$$F = \frac{(R^2_{\text{full}} - R^2_{\text{subset}})/d}{(1 - R^2_{\text{full}})/(n - p - 1)}.$$

That is, if the strength of the fit of the full model (measured by R^2) isn't much larger than that of the subset model, the F-statistic is small, and we do not reject the subset model; if, on the other hand, the difference in R^2 values is large (implying that the fit of the full model is noticeably stronger), we do reject the subset model in favor of the full model.

The F-statistic to test the overall significance of the regression is a special case of this construction (with restriction $\beta_1 = \cdots = \beta_p = 0$), as is each of the individual t-statistics that test the significance of any variable (with restriction $\beta_j = 0$). In the latter case $F_j = t_j^2$.

2.2.2 COLLINEARITY

Recall that the importance of a predictor can be difficult to assess using t-tests when predictors are correlated with each other. A related issue is that of **collinearity** (sometimes somewhat redundantly referred to as **multicollinearity**), which refers to the situation when (some of) the predictors are highly correlated with each other. Predicting variables that are highly correlated with each other can lead to instability in the regression coefficients, and as a result the t-statistics for the variables can be deflated. This can be seen in Figure 2.1. The two plots refer to identical data sets, other than the one data point that is a different color. Dropping the data points down to the (x_1, x_2) plane makes clear the high correlation between the predictors. The estimated regression plane changes from

$$\hat{y} = 9.906 - 2.514x_1 + 6.615x_2$$

in the top plot to

$$\hat{y} = 9.748 + 9.315x_1 - 5.204x_2$$

in the bottom plot; a small change in only one data point causes a major change in the estimated regression function.

Thus, from a practical point of view, collinearity leads to two problems. First, it can happen that the overall F-statistic is significant, yet each of the individual t-statistics is not significant (more generally, the tail probability for the F-test is considerably smaller than those of any of the individual coefficient t-tests). Second, if the data are changed only slightly, the fitted regression coefficients can change dramatically. Note that while collinearity can have a large effect on regression coefficients and associated t-statistics, it does not have a large effect on overall measures of fit like the overall F-test or R^2, since

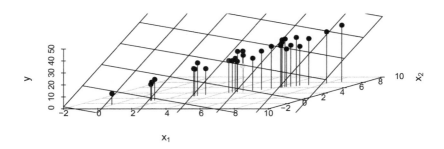

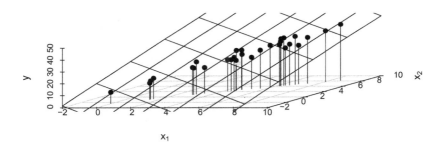

FIGURE 2.1 **Least squares estimation under collinearity. The only change in the data sets is the colored data point. The planes are the estimated least squares fits.**

adding unneeded variables (whether or not they are collinear with predictors already in the model) cannot increase the residual sum of squares (it can only decrease it or leave it roughly the same).

Another problem with collinearity comes from attempting to use a fitted regression model for prediction. As was noted in the previous chapter, simple models tend to forecast better than more complex ones, since they make fewer assumptions about what the future looks like. If a model exhibiting collinearity is used for future prediction, the implicit assumption is that the relationships among the predicting variables, as well as their relationship with the target variable, remain the same in the future. This is less likely to be true if the predicting variables are collinear.

How can collinearity be diagnosed? The two-predictor model

$$y_i = \beta_0 + \beta_1 x_{1i} + \beta_2 x_{2i} + \varepsilon_i$$

TABLE 2.1 **Variance inflation caused by correlation of predictors in a two-predictor model.**

r_{12}	Variance inflation
0.00	1.00
0.50	1.33
0.70	1.96
0.80	2.78
0.90	5.26
0.95	10.26
0.97	16.92
0.99	50.25
0.995	100.00
0.999	500.00

provides some guidance. It can be shown that in this case

$$var(\hat{\beta}_1) = \sigma^2 \left[\sum_{i=1}^{n} x_{1i}^2 (1 - r_{12}^2) \right]^{-1}$$

and

$$var(\hat{\beta}_2) = \sigma^2 \left[\sum_{i=1}^{n} x_{2i}^2 (1 - r_{12}^2) \right]^{-1},$$

where r_{12} is the correlation between x_1 and x_2. Note that as collinearity increases ($r_{12} \to \pm 1$), both variances tend to ∞. This effect is quantified in Table 2.1.

This ratio describes by how much the variances of the estimated slope coefficients are inflated due to observed collinearity relative to when the predictors are uncorrelated. It is clear that when the correlation is high, the variability (and hence the instability) of the estimated slopes can increase dramatically.

A diagnostic to determine this in general is the **variance inflation factor** (VIF) for each predicting variable, which is defined as

$$VIF_j = \frac{1}{1 - R_j^2},$$

where R_j^2 is the R^2 of the regression of the variable x_j on the other predicting variables. VIF_j gives the proportional increase in the variance of $\hat{\beta}_j$ compared to what it would have been if the predicting variables had been uncorrelated. There are no formal cutoffs as to what constitutes a large VIF.

Collinearity is generally not a problem if the observed VIF satisfies

$$VIF < \max\left(10, \frac{1}{1 - R^2_{\text{model}}}\right),$$

where R^2_{model} is the usual R^2 for the regression fit. This means that either the predictors are more related to the target variable than they are to each other, or they are not related to each other very much. In either case coefficient estimates are not very likely to be very unstable, so collinearity is not a problem. If collinearity is present, a simplified model should be considered, but this is only a general guideline; sometimes two (or more) collinear predictors might be needed in order to adequately model the target variable. In the next section we discuss a methodology for judging the adequacy of fitted models and comparing them.

2.3 Methodology

2.3.1 MODEL SELECTION

We saw in Section 2.2.1 that hypothesis tests can be used to compare models. Unfortunately, there are several reasons why such tests are not adequate for the task of choosing among a set of candidate models for the appropriate model to use.

In addition to the effects of correlated predictors on t-tests noted earlier, partial F-tests only can compare models that are nested (that is, where one is a special case of the other). Comparing a model based on $\{x_1, x_3, x_5\}$ to one based on $\{x_2, x_4\}$, for example, is clearly important, but is impossible using these testing methods.

Even ignoring these issues, hypothesis tests don't necessarily address the question a data analyst is most interested in. With a large enough sample, almost any estimated slope will be significantly different from zero, but that doesn't mean that the predictor provides additional *useful* predictive power. Similarly, in small samples, important effects might not be statistically significant at typical levels simply because of insufficient data. That is, there is a clear distinction between statistical significance and practical importance.

In this section we discuss a strategy for determining a "best" model (or more correctly, a set of "best" models) among a larger class of candidate models, using objective measures designed to reflect a predictive point of view. As a first step, it is good to explicitly identify what should *not* be done. In recent years it has become commonplace for databases to be constructed with hundreds (or thousands) of variables and hundreds of thousands (or millions) of observations. It is tempting to avoid issues related to choosing the potential set of candidate models by considering all of the variables as potential predictors in a regression model, limited only by available computing power. This would be a mistake. If too large a set of possible predictors is considered, it is very likely that variables will be identified as important just due to random chance. Since they do not reflect real relationships in the population, models based on them will predict poorly in the future, and interpretations of slope coefficients will just be mistaken justifications of random behavior. This sort of overfitting is known as "data dredging," and is among the most serious dangers when analyzing data.

The set of possible models should ideally be chosen before seeing any data based on as thorough an understanding of the underlying random process as possible. Potential predictors should be justifiable on theoretical grounds if at all possible. This is by necessity at least somewhat subjective, but good basic principles exist. Potential models to consider should be based on the scientific literature and previous relevant experiments. In particular, if a model simply doesn't "make sense," it shouldn't be considered among the possible candidates. That does not mean that modifications and extensions of models that are suggested by the analysis should be ignored (indeed, this is the subject of the next three chapters), but an attempt to keep models grounded in what is already understood about the underlying process is always a good idea.

What do we mean by the (or a) "best" model? As was stated on page 4, there is no "true" model, since any model is only a representation of reality (or equivalently, the true model is too complex to be modeled usefully). Since the goal is not to find the "true" model, but rather to find a model or set of models that best balances fit and simplicity, any strategy used to guide model selection should be consistent with this principle. The goal is to provide a good predictive model that also provides useful descriptions of the process being studied from estimated parameters.

Once a potential set of predictors is chosen, most statistical packages include the capability to produce summary statistics for all possible regression models using those predictors. Such algorithms (often called **best-subsets** algorithms), such as the one described in Furnival and Wilson (1974), do not actually look at all possible models, but rather list statistics for only the models with strongest fits for each number of predictors in the model. This is much less computationally intensive. Such a listing can then be used to determine a set of potential "best" models to consider more closely.

Note that model comparisons are only sensible when based on the same data set. Most statistical packages drop any observations that have missing data in any of the variables in the model. If a data set has missing values scattered over different predictors, the set of observations with complete data will change depending on which variables are in the model being examined, and model comparison measures will not be comparable. One way around this is to only use observations with complete data for all variables under consideration, but this can result in discarding a good deal of available information for any particular model.

2.3.2 EXAMPLE — ESTIMATING HOME PRICES (CONTINUED)

Consider again the home price data examined in Section 1.4. We repeat the regression output from the model based on all of the predictors below:

```
Coefficients:
               Estimate Std.Error t value Pr(>|t|)     VIF
(Intercept)  -7.149e+06 3.820e+06  -1.871 0.065043        .
Bedrooms     -1.229e+04 9.347e+03  -1.315 0.192361  1.262
Bathrooms     5.170e+04 1.309e+04   3.948 0.000171  1.420  ***
Living.area   6.590e+01 1.598e+01   4.124 9.22e-05  1.661  ***
Lot.size     -8.971e-01 4.194e+00  -0.214 0.831197  1.074
Year.built    3.761e+03 1.963e+03   1.916 0.058981  1.242  .
Property.tax  1.476e+00 2.832e+00   0.521 0.603734  1.300
---
Signif. codes:
   0 '***' 0.001 '**' 0.01 '*' 0.05 '.' 0.1 ' ' 1

Residual standard error: 47380 on 78 degrees of freedom
Multiple R-squared: 0.5065,    Adjusted R-squared: 0.4685
F-statistic: 13.34 on 6 and 78 DF,  p-value: 2.416e-10
```

This is identical to the output given earlier, except that variance inflation factor (VIF) values are given for each predictor. It is apparent that there is virtually no collinearity among these predictors (recall that 1 is the minimum possible value of the VIF), which should make model selection more straightforward. The following output summarizes a best-subsets fitting:

Vars	R-Sq	R-Sq(adj)	Mallows Cp	AICc	S	Bedrooms	Bathrooms	Living.area	Lot.size	Year.built	Property.tax
1	35.3	34.6	21.2	1849.9	52576	X					
1	29.4	28.6	30.6	1857.3	54932		X				
1	10.6	9.5	60.3	1877.4	61828						X
2	46.6	45.2	5.5	1835.7	48091	X	X				
2	38.9	37.5	17.5	1847.0	51397		X	X			
2	37.8	36.3	19.3	1848.6	51870	X					X
3	49.4	47.5	3.0	1833.1	47092	X	X	X			
3	48.2	46.3	4.9	1835.0	47635	X	X	X			
3	46.6	44.7	7.3	1837.5	48346	X	X				X
4	50.4	48.0	3.3	1833.3	46885	X	X	X			
4	49.5	47.0	4.7	1834.8	47304	X	X		X		X

```
4   49.4          46.9          5.0 1835.1 47380      X X X X
5   50.6          47.5          5.0 1835.0 47094   X X X     X X
5   50.5          47.3          5.3 1835.2 47162   X X X X X
5   49.6          46.4          6.7 1836.8 47599      X X X X X
6   50.6          46.9          7.0 1836.9 47381   X X X X X X
```

Output of this type provides the tools to choose among candidate models. The output provides summary statistics for the three models with strongest fit for each number of predictors. So, for example, the best one-predictor model is based on Bathrooms, while the second best is based on Living.area; the best two-predictor model is based on Bathrooms and Living.area; and so on. The principle of parsimony noted earlier implies moving down the table as long as the gain in fit is big enough, but no further, thereby encouraging simplicity. A reasonable model selection strategy would not be based on only one possible measure, but looking at all of the measures together, using various guidelines to ultimately focus in on a few models (or only one) that best trade off strength of fit with simplicity, for example as follows:

1. Increase the number of predictors until the R^2 value levels off. Clearly the highest R^2 for a given p cannot be smaller than that for a smaller value of p. If R^2 levels off, that implies that additional variables are not providing much additional fit. In this case the largest R^2 values go from roughly 35% to 47% from $p = 1$ to $p = 2$, which is clearly a large gain in fit, but beyond that more complex models do not provide much additional fit (particularly past $p = 3$). Thus, this guideline suggests choosing either $p = 2$ or $p = 3$.

2. Choose the model that maximizes the adjusted R^2. Recall from equation (1.5) that the adjusted R^2 equals

$$R_a^2 = R^2 - \frac{p}{n - p - 1}\left(1 - R^2\right).$$

It is apparent that R_a^2 explicitly trades off strength of fit (R^2) versus simplicity [the multiplier $p/(n - p - 1)$], and can decrease if predictors that do not add any predictive power are added to a model. Thus, it is reasonable to not complicate a model beyond the point where its adjusted R^2 increases. For these data R_a^2 is maximized at $p = 4$.

The fourth column in the output refers to a criterion called Mallows' C_p (Mallows, 1973). This criterion equals

$$C_p = \frac{\text{Residual SS}_p}{\hat{\sigma}_*^2} - n + 2p + 2,$$

where Residual SS_p is the residual sum of squares for the model being examined, p is the number of predictors in that model, and $\hat{\sigma}_*^2$ is the residual mean square based on using all p^* of the candidate predicting variables. C_p is designed to estimate the expected squared prediction error of a model. Like R_a^2,

C_p explicitly trades off strength of fit versus simplicity, with two differences: it is now small values that are desirable, and the penalty for complexity is stronger, in that the penalty term now multiplies the number of predictors in the model by 2, rather than by 1 (which means that using R_a^2 will tend to lead to more complex models than using C_p will). This suggests another model selection rule:

3. Choose the model that minimizes C_p. In case of tied values, the simplest model (smallest p) would be chosen. In these data, this rule implies choosing $p = 3$.

An additional operational rule for the use of C_p has been suggested. When a particular model contains all of the necessary predictors, the residual mean square for the model should be roughly equal to σ^2. Since the model that includes all of the predictors should also include all of the necessary ones, $\hat{\sigma}_*^2$ should also be roughly equal to σ^2. This implies that if a model includes all of the necessary predictors, then

$$C_p \approx \frac{(n-p-1)\sigma^2}{\sigma^2} - n + 2p + 2 = p + 1.$$

This suggests the following model selection rule:

4. Choose the simplest model such that $C_p \approx p+1$ or smaller. In these data, this rule implies choosing $p = 3$.

A weakness of the C_p criterion is that its value depends on the largest set of candidate predictors (through $\hat{\sigma}_*^2$), which means that adding predictors that provide no predictive power to the set of candidate models can change the choice of best model. A general approach that avoids this is through the use of statistical information. A detailed discussion of the determination of information measures is beyond the scope of this book, but Burnham and Anderson (2002) provides extensive discussion of the topic. The Akaike Information Criterion AIC, introduced by Akaike (1973),

$$AIC = n\log(\hat{\sigma}^2) + n\log[(n-p-1)/n] + 2p + 4, \qquad (2.2)$$

where the $\log(\cdot)$ function refers to natural logs, is such a measure, and it estimates the information lost in approximating the true model by a candidate model. It is clear from (2.2) that minimizing AIC achieves the goal of balancing strength of fit with simplicity, and because of the $2p$ term in the criterion this will result in the choice of similar models as when minimizing C_p. It is well known that AIC has a tendency to lead to overfitting, particularly in small samples. That is, the penalty term in AIC designed to guard against too complicated a model is not strong enough. A modified version of AIC that helps address this problem is the corrected AIC,

$$AIC_c = AIC + \frac{2(p+2)(p+3)}{n-p-3} \qquad (2.3)$$

(Hurvich and Tsai, 1989). Equation (2.3) shows that (especially for small samples) models with fewer parameters will be more strongly preferred when minimizing AIC_c than when minimizing AIC, providing stronger protection against overfitting. In large samples the two criteria are virtually identical, but in small samples, or when considering models with a large number of parameters, AIC_c is the better choice. This suggests the following model selection rule:

5. Choose the model that minimizes AIC_c. In case of tied values, the simplest model (smallest p) would be chosen. In these data, this rule implies choosing $p = 3$, although the AIC_c value for $p = 4$ is virtually identical to that of $p = 3$. Note that the overall level of the AIC_c values is not meaningful, and should not be compared to C_p values or values for other data sets; it is only the value for a model for a given data set relative to the values of others for that data set that matter.

C_p, AIC, and AIC_c have the desirable property that they are *efficient* model selection criteria. This means that in the (realistic) situation where the set of candidate models does not include the "true" model (that is, a good model is just viewed as a useful approximation to reality), as the sample gets larger the error obtained in making predictions using the model chosen using these criteria becomes indistinguishable from the error obtained using the best possible model among all candidate models. That is, in this large-sample predictive sense, it is as if the best approximation was known to the data analyst. Another well-known criterion, the Bayesian Information Criterion BIC [which substitutes $\log(n) \times p$ for $2p$ in (2.2)], does not have this property.

A final way of comparing models is from a directly predictive point of view. Since a rough 95% prediction interval is $\pm 2\hat{\sigma}$, a useful model from a predictive point of view is one with small $\hat{\sigma}$, suggesting choosing a model that has small $\hat{\sigma}$ while still being as simple as possible. That is,

6. Increase the number of predictors until $\hat{\sigma}$ levels off. For these data (S in the output refers to $\hat{\sigma}$) this implies choosing $p = 3$ or $p = 4$.

Taken together, all of these rules imply that the appropriate set of models to consider are those with two, three, or four predictors. Typically the strongest model of each size (which will have highest R^2, highest R_a^2, lowest C_p, lowest AIC_c, and lowest $\hat{\sigma}$, so there is no controversy as to which one is strongest) is examined. The output on page 31 provides summaries for the top three models of each size, in case there are reasons to examine a second- or third-best model (if, for example, a predictor in the best model is difficult or expensive to measure), but here was focus on the best model of each size. First, here is output for the best four-predictor model.

```
Coefficients:
              Estimate Std.Error t value Pr(>|t|)    VIF
(Intercept) -6.852e+06 3.701e+06  -1.852   0.0678        .
Bedrooms    -1.207e+04 9.212e+03  -1.310   0.1940 1.252
Bathrooms    5.303e+04 1.275e+04   4.160 7.94e-05 1.374 ***
Living.area  6.828e+01 1.460e+01   4.676 1.17e-05 1.417 ***
Year.built   3.608e+03 1.898e+03   1.901   0.0609 1.187 .
---
Signif. codes:
  0 '***' 0.001 '**' 0.01 '*' 0.05 '.' 0.1 ' ' 1

Residual standard error: 46890 on 80 degrees of freedom
Multiple R-squared: 0.5044,      Adjusted R-squared: 0.4796
F-statistic: 20.35 on 4 and 80 DF,  p-value: 1.356e-11
```

The t-statistic for number of bedrooms suggests very little evidence that it adds anything useful given the other predictors in the model, so we consider now the best three-predictor model. This happens to be the best four-predictor model with the one statistically insignificant predictor omitted, but this does not have to be the case.

```
Coefficients:
              Estimate Std.Error t value Pr(>|t|)    VIF
(Intercept) -7.653e+06 3.666e+06  -2.087 0.039988        *
Bathrooms    5.223e+04 1.279e+04   4.084 0.000103 1.371 ***
Living.area  6.097e+01 1.355e+01   4.498 2.26e-05 1.210 ***
Year.built   4.001e+03 1.883e+03   2.125 0.036632 1.158 *
---
Signif. codes:
  0 '***' 0.001 '**' 0.01 '*' 0.05 '.' 0.1 ' ' 1

Residual standard error: 47090 on 81 degrees of freedom
Multiple R-squared: 0.4937,      Adjusted R-squared: 0.475
F-statistic: 26.33 on 3 and 81 DF,  p-value: 5.489e-12
```

Each of the predictors is statistically significant at a 0.05 level, and this model recovers virtually all of the available fit ($R^2 = 49.4\%$, while that using all six predictors is $R^2 = 50.6\%$), so this seems to be a reasonable model choice. The estimated slope coefficients are very similar to those from the model using all predictors (which is not surprising given the low collinearity in the data), so the interpretations of the estimated coefficients on page 17 still hold to a large extent. A plot of the residuals versus the fitted values and a normal plot of the residuals (Figure 2.2) look fine, and similar to those for the model on all six predictors in Figure 1.5; plots of the residuals versus each of the predictors in the model are similar to those in Figure 1.6, so they are not repeated here.

Once a "best" model is chosen, it is tempting to use the usual inference tools (such as t-tests and F-tests) to try to explain the process being studied. Unfortunately, doing this while ignoring the model selection process can

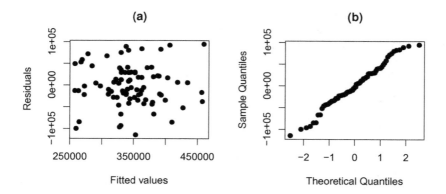

FIGURE 2.2 Residual plots for the home price data using the best three-predictor model. (a) Plot of residuals versus fitted values. (b) Normal plot of the residuals.

lead to problems. Since the model was chosen to be best (in some sense) it will tend to appear stronger than would be expected just by random chance. Conducting inference based on the chosen model as if it was the only one examined ignores an additional source of variability, that of actually choosing the model (model selection based on a different sample from the same population could very well lead to a different chosen "best" model). This is termed **model selection uncertainty**. As a result of ignoring model selection uncertainty, confidence intervals can have lower coverage than the nominal value, hypothesis tests can reject the null too often, and prediction intervals can be too narrow for their nominal coverage.

Identifying and correcting for this uncertainty is a difficult problem, and an active area of research, but there are a few things practitioners can do. First, it is not appropriate to emphasize too strongly the single "best" model; any model that has similar criteria values (such as AIC_c or $\hat{\sigma}$) to those of the best model should be recognized as being one that could easily have been chosen as best based on a different sample from the same population, and any implications of such a model should be viewed as being as valid as those from the best model.

There is a straightforward way to get a sense of the predictive power of a chosen model if enough data are available. This can be evaluated by holding out some data from the analysis (a **holdout** or **validation** sample), applying the selected model from the original data to the holdout sample (based on the previously estimated parameters, not estimates based on the new data), and then examining the predictive performance of the model. If, for example, the standard deviation of the errors from this prediction is not very different from the standard error of the estimate in the original regression, chances are that making inferences based on the chosen model will not be misleading. Similarly, if a (say) 95% prediction interval does not include roughly 95% of

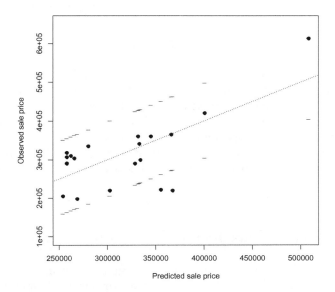

FIGURE 2.3 Plot of observed versus predicted house sale price values of validation sample, with pointwise 95% prediction interval limits superimposed. The dotted line corresponds to equality of observed values and predictions.

the new observations, that indicates poorer-than-expected predictive performance on new data.

Figure 2.3 illustrates a validation of the three-predictor housing price model on a holdout sample of 20 houses. The figure is a plot of the observed versus predicted prices, with pointwise 95% prediction interval limits superimposed. The intervals contain 90% of the prices (18 of 20), and the average predictive error on the new houses is only \$3429 (compared to an average observed price of more than \$313,000), not suggesting the presence of any forecasting bias in the model. Two of the houses, however, have sale prices well below what would have been expected (more than \$130,000 lower than expected), and this is reflected in a much higher standard deviation (\$66,308) of the predictive errors than $\hat{\sigma} = \$47,090$ from the fitted regression. If the two outlying houses are omitted the standard deviation of the predictive errors is much smaller (\$49,515), suggesting that while the fitted model's predictive performance for most houses is in line with its performance on the original sample, there are indications that it might not predict well for the occasional unusual house.

If validating the model on new data this way is not possible, a simple adjustment that is helpful is to estimate the variance of the errors as

$$\tilde{\sigma}^2 = \frac{\sum_{i=1}^{n}(y_i - \hat{y}_i)^2}{n - p^* - 1},\tag{2.4}$$

where $\hat{y}$ is based on the chosen "best" model, and p^* is the number of predictors in the most complex model examined, in the sense of most predictors (Ye, 1998). Clearly if very complex models are included among the set of candidate models, $\tilde{\sigma}$ can be much larger than the standard error of the estimate from the chosen model, with correspondingly wider prediction intervals. This reinforces the benefit of limiting the set of candidate models (and the complexity of the models in that set) from the start. In this case $\tilde{\sigma} = \$47,987$, so the effect is not that pronounced.

2.4   Indicator Variables and Modeling Interactions

It is not unusual for the observations in a sample to fall into two distinct subgroups; for example, people are either male or female. It might be that group membership has no relationship with the target variable (given other predictors); such a **pooled model** ignores the grouping and pools the two groups together.

On the other hand, it is clearly possible that group membership is predictive for the target variable (for example, expected salaries differing for men and women given other control variables could indicate gender discrimination). Such effects can be explored easily using an **indicator variable**, which takes on the value 0 for one group and 1 for the other (such variables are sometimes called **dummy variables** or $0/1$ **variables**). The model takes the form

$$y_i = \beta_0 + \beta_1 x_{1i} + \cdots + \beta_{p-1} x_{p-1,i} + \beta_p \mathcal{I}_i + \varepsilon_i,$$

where $\mathcal{I}_i$ is an indicator variable with value 1 if the observation is a member of group and 0 otherwise. The usual interpretation of the slope still applies: β_p is the expected change in y associated with a one-unit change in $\mathcal{I}$ holding all else fixed. Since $\mathcal{I}$ only takes on the values 0 or 1, this is equivalent to saying that the expected target is β_p higher for group members ($\mathcal{I} = 1$) than nonmembers ($\mathcal{I} = 0$), holding all else fixed. This has the appealing interpretation of fitting a **constant shift model**, where the regression relationships for group members and nonmembers are identical, other than being shifted up or down; that is,

$$y_i = \beta_0 + \beta_1 x_{1i} + \cdots + \beta_{p-1} x_{p-1,i} + \varepsilon_i$$

for nonmembers and

$$y_i = \beta_0 + \beta_p + \beta_1 x_{1i} + \cdots + \beta_{p-1} x_{p-1,i} + \varepsilon_i$$

for members. The t-test for whether $\beta_p = 0$ is thus a test of whether a constant shift model (two parallel regression lines, planes, or hyperplanes) is a significant improvement over a pooled model (one common regression line, plane, or hyperplane).

Would two *different* regression relationships be better still? Say there is only one numerical predictor x; the full model that allows for two different regression lines is

$$y_i = \beta_{00} + \beta_{10} x_{1i} + \varepsilon_i$$

for nonmembers ($\mathcal{I} = 0$), and

$$y_i = \beta_{01} + \beta_{11} x_{1i} + \varepsilon_i$$

for members ($\mathcal{I} = 1$). The pooled model and the constant shift model can be made to be special cases of the full model, by creating a new variable that is the product of x and $\mathcal{I}$. A regression model that includes this variable,

$$y_i = \beta_0 + \beta_1 x_{1i} + \beta_2 \mathcal{I}_i + \beta_3 x_{1i} \mathcal{I}_i + \varepsilon_i,$$

corresponds to the two different regression lines

$$y_i = \beta_0 + \beta_1 x_{1i} + \varepsilon_i$$

for nonmembers (since $\mathcal{I} = 0$), implying $\beta_{00} = \beta_0$ and $\beta_{10} = \beta_1$ above, and

$$y_i = \beta_0 + \beta_1 x_{1i} + \beta_2 + \beta_3 x_{1i} + \varepsilon_i$$
$$= (\beta_0 + \beta_2) + (\beta_1 + \beta_3) x_{1i} + \varepsilon_i$$

for members (since $\mathcal{I} = 1$), implying $\beta_{01} = \beta_0 + \beta_2$ and $\beta_{11} = \beta_1 + \beta_3$ above.

The t-test for the slope of the product variable ($\beta_3 = 0$) is a test of whether the full model (two different regression lines) is significantly better than the constant shift model (two parallel regression lines); that is, it is a test of parallelism. The restriction $\beta_2 = \beta_3 = 0$ defines the pooled model as a special case of the full model, so the partial F-statistic based on (2.1),

$$F = \frac{(\text{Residual SS}_{\text{pooled}} - \text{Residual SS}_{\text{full}})/2}{\text{Residual SS}_{\text{full}}/(n - 4)}$$

on $(2, n - 4)$ degrees of freedom, provides a test comparing the pooled model to the full model. This test is often called the **Chow test** (Chow, 1960) in the economics literature.

These constructions can be easily generalized to multiple predictors, with different variations of models obtainable. For example, a regression model with unequal slopes for some predictors and equal slopes for others is fit by including products of the indicator and the predictor for the ones with different slopes and not including them for the predictors with equal slopes. Appropriate t- and F-tests can then be constructed to make particular comparisons of models.

A reasonable question to ask at this point is "Why bother to fit the full model? Isn't it just the same as fitting two separate regressions on the two groups?" The answer is no. The full model fit above assumes that the variance of the errors is the same (the constant variance assumption), while fitting two separate regressions allows the variances to be different. The fitted slope coefficients from the full model will, however, be identical to those from two separate fits. What is gained by analyzing the data this way is the comparison of versions of pooled, constant shift, and full models based on group membership, including different slopes for some variables and equal slopes for others, something that is not possible if separate regressions are fit to the two groups.

Another way of saying that the relationship between a predictor and the target is different for members of the two different groups is that there is an **interaction effect** between the predictor and group membership on the target. Social scientists would say that the grouping has a *moderating* effect on the relationship between the predictor and the target. The fact that in the case of a grouping variable the interaction can be fit by multiplying the two variables together has led to a practice that is common in some fields: to try to represent *any* interaction between variables (that is, any situation where the relationship between a predictor and the target is different for different values of another predictor) by multiplying them together. Unfortunately, this is not a very reasonable way to think about interactions for numerical predictors, since there are many ways that the effect of one variable on the target can differ depending on the value of another that have nothing to do with product functions.

2.4.1 EXAMPLE — ELECTRONIC VOTING AND THE 2004 PRESIDENTIAL ELECTION

The 2000 U.S. presidential election matching Republican George W. Bush against Democrat Al Gore attracted worldwide attention because of its close and controversial results, particularly in the state of Florida. The 2004 election, pitting the incumbent Bush against John Kerry, is less discussed, but was also controversial, in part because of the introduction of electronic voting machines in some polling places across the country (such machines were introduced in part because of the irregularities in paper balloting that occurred in Florida in the 2000 election). Some of the manufacturers of electronic voting machines were strong supporters of President Bush, and this, along with the fact that the machines did not produce a paper trail, led to speculation about whether the machines could be manipulated to favor one candidate over the other.

This analysis is based on data from Hout et al. (2004) (see also Theus and Urbanek, 2009). The observations are the 67 counties of Florida. Although this is not a sample of Florida counties (it is actually a census of all of them), these counties can be considered a sample of all of the counties in the country, making inferences drawn about the larger population of counties based on this set of counties meaningful. The target variable is the change in the

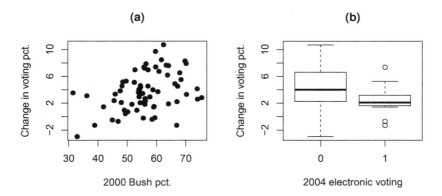

FIGURE 2.4 Plots for the 2004 election data. (a) Plot of percentage change in Bush vote versus 2000 Bush vote. (b) Side-by-side boxplots of percentage change in Bush vote by whether or not the county employed electronic voting in 2004.

percentage of votes cast for Bush from 2000 to 2004 (a positive number meaning a higher percentage in 2004). We start with the simple regression model relating the change in Bush percentage to the percentage of votes Bush took in 2000, with corresponding scatter plot given in the left plot of Figure 2.4. It can be seen that most of the changes are positive, reflecting that Bush carried the state by more than 380,000 votes in 2004, compared with the very close result (a 537 vote margin) in 2000.

```
Coefficients:
                Estimate Std. Error t value Pr(>|t|)
(Intercept)     -2.9968    2.0253    -1.480  0.14379
Bush.pct.2000    0.1190    0.0355     3.352  0.00134 **
---
Signif. codes:
  0 '***' 0.001 '**' 0.01 '*' 0.05 '.' 0.1 ' ' 1

Residual standard error: 2.693 on 65 degrees of freedom
Multiple R-squared: 0.1474,    Adjusted R-squared: 0.1343
F-statistic: 11.24 on 1 and 65 DF,  p-value: 0.00134
```

There is a weak, but statistically significant, relationship between 2000 Bush vote and the change in vote to 2004, with counties that went more strongly for Bush in 2000 gaining more in 2004. The constant shift model now adds an indicator variable for whether a county used electronic voting in 2004. The side-by-side boxplots in the right plot in Figure 2.4 show that overall the 15 counties that used electronic voting had smaller gains for Bush than the 52 that did not, but that of course does not take the 2000 Bush vote

into account. There are also signs of nonconstant variance, as the variability
is smaller among the counties that used electronic voting.

```
Coefficients:
                Estimate Std. Error t value Pr(>|t|)    VIF
(Intercept)     -2.12713    2.10315  -1.011  0.31563
Bush.pct.2000    0.10804    0.03609   2.994  0.00391 1.049 **
e.Voting        -1.12840    0.80218  -1.407  0.16437 1.049
---

Signif. codes:
  0 '***' 0.001 '**' 0.01 '*' 0.05 '.' 0.1 ' ' 1

Residual standard error: 2.672 on 64 degrees of freedom
Multiple R-squared: 0.173,       Adjusted R-squared: 0.1471
F-statistic: 6.692 on 2 and 64 DF,  p-value: 0.002295
```

It can be seen that there is only weak (if any) evidence that the constant
shift model provides improved performance over the pooled model. This does
not mean that electronic voting is irrelevant, however, as it could be that two
separate (unrestricted) lines are preferred.

```
Coefficients:
                Estimate Std.Error t value Pr(>|t|)    VIF
(Intercept)     -5.23862   2.35084  -2.228 0.029431          *
Bush.pct.2000    0.16228   0.04051   4.006 0.000166  1.44 ***
e.Voting         9.67236   4.26530   2.268 0.026787 32.26 *
Bush.2000
  X e.Voting    -0.20051   0.07789  -2.574 0.012403 31.10 *
---
Signif. codes:
  0 '***' 0.001 '**' 0.01 '*' 0.05 '.' 0.1 ' ' 1

Residual standard error: 2.562 on 63 degrees of freedom
Multiple R-squared: 0.2517,      Adjusted R-squared: 0.2161
F-statistic: 7.063 on 3 and 63 DF,  p-value: 0.0003626
```

The t-test for the product variable indicates that the model with two unrestricted lines is preferred over the model with two parallel lines. A partial F-test comparing this model to the pooled model, which is $F = 4.39$
($p = .016$), also supports two distinct lines,

$$\text{Change.in.Bush.pct} = -5.239 + .162 \times \text{Bush.pct.2000}$$

for counties that did not use electronic voting in 2004, and

$$\text{Change.in.Bush.pct} = 4.434 - .038 \times \text{Bush.pct.2000}$$

for counties that did use electronic voting. This is represented in Figure 2.5.
This relationship implies that in counties that did not use electronic voting

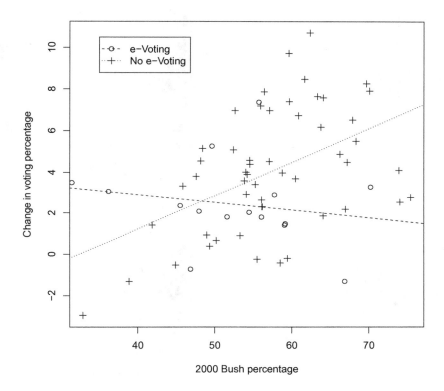

FIGURE 2.5 Regression lines for election data separated by whether the county used electronic voting in 2004.

the more Republican a county was in 2000 the larger the gain for Bush in 2004, while in counties with electronic voting the opposite pattern held true.

As can be seen from the VIFs, the predictor and the product variable are collinear. This isn't very surprising, since one is a function of the other, and such collinearity is more likely to occur if one of the subgroups is much larger than the other, or if group membership is related to the level or variability of the predictor variable. Given that using the product variable is just a computational construction that allows the fitting of two separate regression lines, this is not a problem in this context.

This model is probably underspecified, as it does not include control variables that would be expected to be related to voting percentage. Figure 2.6 gives scatter plots of the percentage change in Bush votes versus (a) the total county voter turnouts in 2000 and (b) 2004, (c) median income, and (d) percentage of the voters being Hispanic. None of the marginal relationships

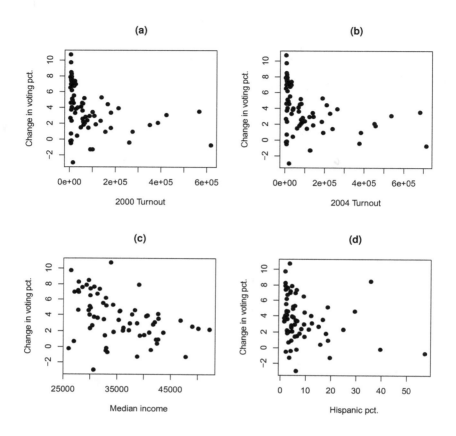

FIGURE 2.6 Plots for the 2004 election data. (a) Plot of percentage change in Bush vote versus 2000 voter turnout. (b) Plot of percentage change in Bush vote versus 2004 voter turnout. (c) Plot of percentage change in Bush vote versus median income. (d) Plot of percentage change in Bush vote versus percentage Hispanic voters.

are very strong, but in the multiple regression summarized below, median income does seem to add important predictive power without changing the previous relationships between change in Bush voting percentage and 2000 Bush percentage very much.

```
Coefficients:
                    Estimate Std.Error   t val P(>|t|)      VIF
(Intercept)        1.166e+00 2.55e+00    0.46  0.650
Bush.pct.2000      1.639e-01 3.69e-02    4.45  3.9e-5     1.55  ***
e.Voting           1.426e+01 4.84e+00    2.95  0.005     54.08  **
Bush.2000
   X e.Voting     -2.545e-01 8.47e-02   -3.01  0.004     47.91  **
Vote.turn.2000    -5.957e-06 3.10e-05   -0.19  0.848    210.66
```

```
Vote.turn.2004  1.413e-06   2.49e-05    0.06   0.955 205.81
Median.income  -1.745e-04   5.61e-05   -3.11   0.003   1.66 **
Hispan.pop.pct -4.127e-02   3.18e-02   -1.30   0.200   1.32
---
Signif. codes:
  0 '***' 0.001 '**' 0.01 '*' 0.05 '.' 0.1 ' ' 1

Residual standard error: 2.244 on 59 degrees of freedom
Multiple R-squared: 0.4624,    Adjusted R-squared: 0.3986
F-statistic:  7.25 on 7 and 59 DF,  p-value: 2.936e-06
```

We could consider simplifying the model here, but often researchers prefer to not remove control variables, even if they do not add to the fit, so that they can be sure that the potential effect is accounted for. This is generally not unreasonable if collinearity is not a problem, but control variables that do not provide additional significant predictive power, but are collinear with the variables that are of direct interest, might be worth removing so they don't obscure the relationships involving the more important variables. In these data the two voter turnout variables are (not surprisingly) highly collinear, but a potential simplification to consider (particularly given that the target variable is the change in Bush voting percentage from 2000 to 2004) is to consider the change in voter turnout as a predictor (the fact that the estimated slope coefficients for 2000 and 2004 voter turnout are of opposite signs and not very different also supports this idea). The model using change in voter turnout is a subset of the model using 2000 and 2004 voter turnout separately (corresponding to restriction $\beta_{2004} = -\beta_{2000}$), so the two models can be compared using a partial F-test. As can be seen below, the fit of the simpler model is similar to that of the more complicated one, collinearity is no longer a problem, and it turns out that the partial F-test ($F = 0.43$, $p = .516$) supports that the simpler model fits well enough compared to the more complicated model to be preferred (although voter turnout is still apparently not important).

```
Coefficients:
                Estimate Std.Error  t val P(>|t|)      VIF
(Intercept)    1.157e+00 2.54e+00    0.46   0.651
Bush.pct.2000  1.633e-01 3.67e-02    4.46 3.7e-05   1.55 ***
e.Voting       1.272e+01 4.20e+00    3.03   0.004 41.25 **
Bush.2000
  X e.Voting  -2.297e-01 7.53e-02   -3.05   0.003 38.25 **
Change.turnout -1.223e-05 1.36e-05   -0.90   0.370   2.44
Median.income  -1.718e-04 5.57e-05   -3.08   0.003   1.65 **
Hispan.pop.pct -4.892e-02 2.94e-02   -1.66   0.102   1.14
---
Signif. codes:
  0 '***' 0.001 '**' 0.01 '*' 0.05 '.' 0.1 ' ' 1

Residual standard error: 2.233 on 60 degrees of freedom
Multiple R-squared: 0.4585,    Adjusted R-squared: 0.4044
F-statistic: 8.468 on 6 and 60 DF,  p-value: 1.145e-06
```

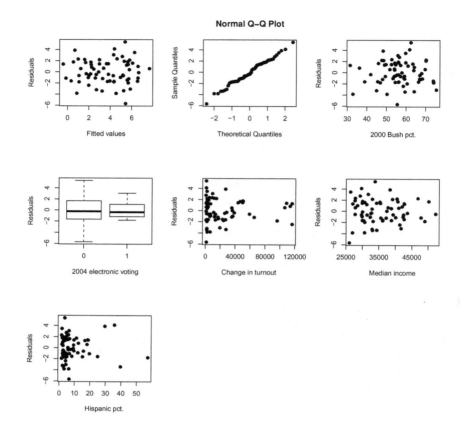

FIGURE 2.7   Residual plots for the 2004 election data.

Residual plots given in Figure 2.7 do not indicate any obvious problems, although the potential nonconstant variance related to whether a county used electronic voting or not noted in Figure 2.4 is still indicated. We will not address that issue here, but correction of nonconstant variance related to subgroups in the data will be discussed in Section 6.3.3.

2.5   Summary

In this chapter we have discussed various issues related to model building and model selection. Such methods are important because both underfitting (not including variables that are needed) and overfitting (including variables that are not needed) lead to problems in interpreting the results of regression analyses and making predictions using fitted regression models. Hypothesis tests provide one tool for model building through formal comparisons of models. If one model is a special case of another, defined through a linear restriction, then a partial F-statistic provides a test of whether the more complex model provides significantly more predictive power than does the simpler one. One important example of a partial F-test is the standard t-test for the significance of a slope coefficient. Another important use of partial F-tests is in the construction of models for data where observations fall into two distinct subgroups that allow for common (pooled) relationships over groups, constant shift relationships that differ only in level but not in slopes, and completely distinct and different relationships across groups.

While useful, hypothesis tests do not provide a complete tool for model building. The problem is that a hypothesis test does not necessarily answer the question that is of primary importance to a data analyst. The t-test for a particular slope coefficient tests whether a variable adds predictive power given the other variables in the model, but if predictors are collinear it could be that none add anything given the others, while separately still being very important. A related problem is that collinearity can lead to great instability in regression coefficients and t-tests, making results difficult to interpret. Hypothesis tests also do not distinguish between statistical significance (whether or not a true coefficient is exactly zero) from practical importance (whether or not a model provides the ability for an analyst to make important discoveries in the context of how a model is used in practice).

These considerations open up a broader spectrum of tools for model building than just hypothesis tests. Best-subsets regression algorithms allow for the quick summarization of hundreds or even thousands of potential regression models. The underlying principle of these summaries is the principle of parsimony, which implies the tradeoff of strength of fit versus simplicity: that a model should only be as complex as it needs to be. Measures such as the adjusted R^2, C_p, and AIC_c explicitly provide this tradeoff, and are useful tools in helping to decide when a simpler model is preferred over a more complicated one. An effective model selection strategy uses these measures, as well as hypothesis tests and estimated prediction intervals, to suggest a set of potential "best" models, which can then be considered further. In doing so, it is important to remember that the variability that comes from model selection itself (model selection uncertainty) means that it is likely that several models actually provide descriptions of the underlying population process that are equally valid. One way of assessing the effects of this type of uncertainty is to keep some of the observed data aside as a holdout sample, and then validate the chosen fitted model(s) on that held out data.

Although best-subsets algorithms and modern computing power have made automatic model selection more feasible than it once was, they are still limited computationally to a maximum of roughly 30 predictors. In recent years it has become more common for a data analyst to be faced with data sets with hundreds or thousands of predictors, making such methods infeasible. Recent work has focused on alternatives to least squares called **regularization methods**, which can be viewed as effectively variable selectors, and are feasible for very large numbers of predictors. See Bühlmann and van de Geer (2011) for detailed discussion of such methods.

KEY TERMS

AIC_c: A modified version of the Akaike Information Criterion (AIC) that guards against overfitting in small samples. It is used to compare models when performing model selection.

Best-subsets regression: A procedure that generates the best-fitting models for each number of predictors in the model.

Chow test: A statistical (partial F-)test for determining whether a single regression model can be used to describe the regression relationships when two groups are present in the data.

Collinearity: When predictor variables in a regression fit are highly correlated with each other.

Constant shift model: Regression models that have different intercepts but the same slope coefficients for the predicting variables for different groups in the data.

Indicator variable: A variable that takes on the values 0 or 1, indicating whether a particular observation belongs to a certain group or not.

Interaction effect: When the relationship between a predictor and the target variable differs depending on the group in which an observation falls.

Linear restriction: A linear condition on the regression coefficients that defines a special case (subset) of a larger unrestricted model.

Mallows' C_p: A criterion used for comparing several competing models to each other. It is designed to estimate the expected squared prediction error of a model.

Model selection uncertainty: The variability in results that comes from the fact that model selection is an iterative process, arrived at after examination of several models, and therefore the final model chosen is dependent on the particular sample drawn from the population. Significance levels, confidence intervals, etc., are not exact, as they depend on a chosen model that is itself random. This should be recognized when interpreting results.

Overfitting: Including redundant or noninformative predictors in a fitted regression.

Partial F-test: F-test used to compare the fit of an unrestricted model to that of a restricted model (defined by a linear restriction), in order to see if the restricted model is adequate to describe the relationship in the data.

Pooled model: A single model fit to the data that ignores group classification.

Underfitting: Omitting informative essential variables in a fitted regression.

Variance inflation factor: A statistic giving the proportional increase in the variance of the sample regression coefficient for a particular predictor due to the linear association of the predictor with other predictors.

PART TWO

Addressing Violations of Assumptions

CHAPTER THREE

Diagnostics for Unusual Observations

3.1 Introduction

As is true of all statistical methodologies, linear regression analysis can be a very effective way to model data as long as the assumptions being made are true, but if they are violated least squares can potentially lead to misleading results. The residual plots discussed in Section 1.3.5 are important tools to check these assumptions, but their flexibility is both a strength and a weakness. The plots can be examined for evidence of violations of assumptions without requiring specification of the exact form of the violations, but the subjective nature of such examination can easily lead to different data analysts having different impressions of the validity of the underlying assumptions. Plots also by definition can only provide two-dimensional views of a multivariate regression relationship.

    In this chapter (and several others to follow) we describe other tools that can be used to identify and address potential problems with the application of linear least squares estimation to regression problems. This chapter discusses

Handbook of Regression Analysis. By Samprit Chatterjee and Jeffrey S. Simonoff
Copyright © 2013 John Wiley & Sons, Inc.

diagnostics for the identification of unusual observations. We will describe both graphical and more formal uses of these diagnostics, but will not emphasize tests of significance.

3.2   Concepts and Background Material

There are several reasons why it is important to identify unusual observations.

1. Unusual observations are sometimes simply mistakes that arise from incorrect entry or faulty measurement of numerical values. Obviously these should be corrected if possible.

2. It is often the case that a great deal of information can come from examination of unusual observations. It is possible that the reason that an observation apparently has a different relationship between the response and the predictor(s) from the one that is typical for the data is that it is different in a fundamental way from the other observations, such as having been measured under different conditions. In such a circumstance it could be that the observation should have never been included in the sample at all, or perhaps the regression model could potentially be enriched to account for these different conditions through additional predictors, aiding in estimation not only for that observation but others as well. Another possibility is that an observation has a very different set of predictor values from what is typical for the bulk of the data; this could suggest sampling more observations with similar values of those predictors.

3. It is important that all of the observations in a sample have similar influence on a fitted model. It is not desirable that just a few of the observations have a strong influence on the fitted regression. The summary of the relationship between the response and the predictor(s) should be based on the bulk of the data, and not just on a small subset of it. This should hold for not only the estimated regression coefficients, but also for measures of its strength, and any variable selection that might be done. *All* of these measures are potentially affected by unusual observations, and the presence of such observations can lead to a misleading model for the data if they are ignored.

It is worth saying a little more about the third reason given above. In situations where an unusual observation is not obviously "wrong" (that is, a point is not unusual because of a transcription error), it is sometimes argued that it is not appropriate to omit unusual observations from a data set, because all of the observations in the data are "legitimate." The argument is to keep the data "as they really are." This is a fundamentally incorrect attitude, as it ignores the key goal of any statistical model, which is to describe as accurately as possible the underlying process driving the bulk of the data. Consider Figure 3.1. The fitted regression line that is based on all of the data

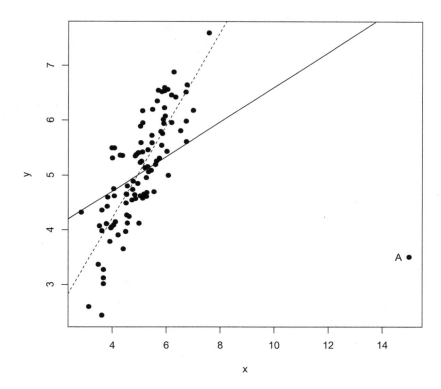

FIGURE 3.1 **Scatter plot of data with an unusual observation (A), with the fitted regression line with (solid line) and without (dashed line) the unusual observation
included.**

(the solid line) is obviously an extremely poor representation of what is going
on in the data — it does not in any way describe the data "as they really are,"
because of the deficiencies of least squares regression modeling (and its sensitivity to unusual observations). A regression method that is insensitive to
unusual observations (a so-called **robust** method) would be affected much less
by the unusual observation, resulting in a fitted line similar to that obtained
when omitting observation A (the dashed line). This is the correct summary
of the data, but the role of A should be noted and reported in the analysis.
That is, the issue is *not* that there is something "wrong" with the data point;
rather, the issue is that there is something wrong with least squares regression
in its sensitivity to unusual observations. For this reason, the least squares
fit with unusual observations omitted should always be examined. If it is noticeably different from that with the unusual observations included it should

be reported, as it is likely to be a better representation of the underlying relationship. When this is done, however, the points omitted must be identified and discussed as part of the summary of the results.

A data analyst should always examine diagnostics and report on unusual observations. This can be done by printing them out, or (particularly if the sample is large) displaying them in observation order (a so-called **index plot**) to help make identification easier. Such a plot also has the advantage of highlighting relative values of diagnostics (compared to the others in the sample), since an observation that is very unusual compared to the sample as a whole is worth investigating further even if it is not objectively "very unusual" based on an arbitrary cutoff value.

3.3   Methodology

3.3.1 RESIDUALS AND OUTLIERS

An **outlier** is an observation with a response value y_i that is unusual relative to its expected value. Since the fitted value $\hat{y}_i$ is the best available estimate of the expected response for the ith observation, it is natural that the residual $e_i = y_i - \hat{y}_i$ should be the key statistic used to evaluate if an observation is an outlier. An outlier is an observation with a large absolute residual (note that residuals can be positive or negative). The issue is then to determine what is meant by "large." As was noted in equation (1.4), the residuals satisfy

$$\mathbf{e} = (I - H)\mathbf{y},$$

where $H = X(X'X)^{-1}X'$ is the hat matrix. Since the variance of y_i satisfies $V(y_i) = \sigma^2$, straightforward algebra shows that

$$V(\mathbf{e}) = (I - H)\sigma^2,$$

or for an individual residual,

$$V(e_i) = \sigma^2(1 - h_{ii}), \tag{3.1}$$

where h_{ii} is the ith diagonal element of the hat matrix. Thus, scaling the residual by its standard deviation

$$e_i^* = \frac{e_i}{\sigma\sqrt{1 - h_{ii}}} \tag{3.2}$$

results in a **standardized residual** that has mean 0 and standard deviation 1. Since the errors are assumed to be normally distributed, the residuals will also be normally distributed, implying that the usual rules of thumb for normally distributed random variables can be applied. For example, since only roughly 1% of the sample from a normal distribution is expected to be outside ±2.5, standardized residuals outside ± 2.5 can be flagged as potentially outlying and examined further.

The standardized residual (3.2) depends on the unknown σ, so actual calculation requires an estimate of σ. The standard approach is to use the standard error of the estimate $\hat{\sigma}$, the square root of the residual mean square (1.6); this is sometimes called the **internally studentized residual**, but is usually just referred to as the standardized residual. An alternative approach is to base the estimate of σ on all of the observations except the ith one when determining the ith residual, so the point will not affect the estimate of σ if it is in fact an outlier; this is called the **externally studentized residual**

$$\tilde{e}_i = \frac{e_i}{\hat{\sigma}_{(i)}\sqrt{1 - h_{ii}}},$$

where $\hat{\sigma}_{(i)}$ is the standard error of the estimate based on the model omitting the ith observation. It can be shown that

$$\tilde{e}_i = e_i^* \sqrt{\frac{n - p - 2}{n - p - 1 - e_i^{*2}}}.$$

The distribution of the externally studentized residuals is simpler than that of the internally studentized (standardized) residuals (they follow a t_{n-p-2}-distribution), and will be larger in absolute value for outlying points, but since one is a monotonic function of the other, plots using the two types of residuals will be similar in appearance. Note that this is also true for the types of residual plots discussed in Section 1.3.5; although the appearance of the plots will be similar no matter which version of the residuals are plotted, given the standard-normal scale on which they are measured, there is no reason not to use some form of standardized residual in all residual plots.

3.3.2 LEVERAGE POINTS

One of the reasons observation A in Figure 3.1 had such a strong effect on the fitted regression is that its value for the predictor was very different from that of the other predictor values. This isolation in the x-space tended to draw the fitted regression line towards it. This can be made more formal. As noted in equation (1.3), the fitted values satisfy $\hat{\mathbf{y}} = H\mathbf{y}$, where H is the hat matrix. This can be written out explicitly for the ith fitted value as

$$\hat{y}_i = h_{i1}y_1 + h_{i2}y_2 + \cdots + h_{ii}y_i + \cdots + h_{in}y_n. \tag{3.3}$$

Thus, the ith diagonal element of the matrix, h_{ii}, represents the potential effect that the ith observed value y_i can have on the ith fitted value $\hat{y}_i$. Since leverage points, by being isolated in the X-space, draw the regression line (or plane or hyperplane) towards them, h_{ii} is an algebraic reflection of the tendency that an observation has to draw the line towards it. The connection between large values of h_{ii} and an unusual position in X-space is particularly clear in the case of simple regression, where

$$h_{ii} = \frac{1}{n} + \frac{(x_i - \overline{X})^2}{\sum_j (x_j - \overline{X})^2};$$

the farther x_i is from the center of the data (as measured by the sample mean of the x's), the higher the leverage.

It can be shown that $0 < h_{ii} < 1$ for all i, and the sum of the n leverage values equals $p + 1$, where p is the number of predicting variables in the regression. That is, the average leverage value is $(p + 1)/n$. A good guideline for what constitutes a large leverage value is $(2.5)(p+1)/n$. Cases with values greater than that should be investigated as possible leverage points.

Based on equation (3.1), it is easy to see that leverage points have residuals with less variability than residuals from non-leverage points (since h_{ii} is closer to 1, resulting in a smaller variance of the residual). This is not surprising; since a leverage point is characterized by a fitted value close to the observed target value (that is, it tends to pull the fitted regression towards it), its residual is likely to be closer to zero. Another way to see this is from the fact that

$$h_{ii} + \frac{e_i^2}{\hat{\sigma}^2(n - p - 1)} \leq 1,$$

which shows that as h_{ii} gets closer to 1, $|e_i|$ gets closer to 0.

3.3.3 INFLUENTIAL POINTS AND COOK'S DISTANCE

As described in the previous section, the idea of leverage is all about the *potential* for an observation to have a large effect on a fitted regression; if the observation does not have an unusual response value, it is possible that drawing the regression towards it will not change the estimated coefficients very much (or at all). Figure 3.2 gives two examples of this pattern. In the top plot the unusual point is a leverage point, but falls almost directly on the line implied by the rest of the data; that is, it is not an outlier. If this point is omitted the fitted regression line will change very little, so in the sense of effect on the estimated coefficients the point is not influential. Note that the deletion may have other effects; for example, the R^2 and overall F-statistic with the point included would probably be noticeably higher than those with the point omitted, as the unusual point increases the total sum of squares $\sum(y_i - \overline{Y})^2$ without increasing the residual sum of squares $\sum(y_i - \hat{y}_i)^2$. In the bottom plot the unusual point is an outlier but not a leverage point. As equation (3.3) shows, since h_{ii} would be relatively small for this point, the observed y_i has little effect on the fitted $\hat{y}_i$, so omitting it changes the fitted regression very little.

Given these different notions of influence, it is not surprising that there are many measures of the influence of an observation in regression analysis.

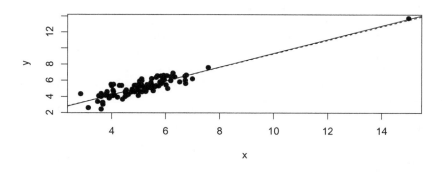

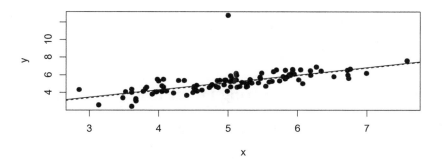

FIGURE 3.2 Scatter plots of data with an unusual observation, with the fitted regression line with (solid line) and without (dashed line) the unusual observation included.

The most widely used measure is the one proposed by Cook (1977), which measures the change in the fitted regression coefficients if a case were dropped from the regression, relative to the inherent variability of the coefficient estimates themselves. **Cook's distance** D is also equivalent to the change in the predicted values from the full data and the fitted value obtained by deleting the observation. Cook's distance combines the notions of outlyingness and leverage in an appealing way, since

$$D_i = \frac{(e_i^*)^2 h_{ii}}{(p+1)(1-h_{ii})}. \tag{3.4}$$

Observations that are outliers (with large absolute standardized residual $|e_i^*|$) or leverage points (with large h_{ii}) are potentially influential, and points that are both (so-called "bad" leverage points) are the most influential. A value of

Cook's D over 1 or so should be flagged. These points should be examined further.

It is important to remember that Cook's D only measures one particular form of influence, and shouldn't be viewed as the final judge of whether or not a point is worth investigating. For example, outliers that have low values of Cook's D can still have a large effect on hypothesis tests, R^2, the standard error of the estimate, and so on. Other measures of influence have been proposed that focus on such notions of influence; see Belsley et al. (1980) and Chatterjee and Hadi (1988) for more discussion.

It is worth noting a weakness of *all* of these diagnostics. Specifically, they are all sensitive to the so-called **masking effect**. This occurs when several unusual observations are all in the same region of the (X, y) space. When this happens, the diagnostics, which all focus on changes in the regression when a *single* point is deleted, fail, since the presence of the other nearby unusual observations means that the fitted regression changes very little if one is omitted. The problem of multiple outliers in regression is a topic of ongoing research, and typically involves defining a "clean" subset of the sample to which potentially outlying observations are compared; see, for example, Hadi and Simonoff (1993) and Atkinson and Riani (2000).

3.4 Example — Estimating Home Prices (continued)

Consider again the home price data examined in Chapters 1 and 2. Regression diagnostics for the chosen model on page 35 are given as index plots in Figure 3.3. The guidelines given earlier for flagging unusual values are given using dotted lines, although this is not given in the Cook's distance plot as the largest value is not close to 1.

None of the points are flagged as outliers or influential points (according to Cook's distance), but there are five leverage points flagged as unusual. The scatter plots given in Figure 1.4 on page 17 show that these correspond to the only two houses with living area greater than 2500 square feet (each having living area more than 2900 square feet), and the only three houses built after 1955 (each being built in 1961 or 1962). It is possible that the underlying relationship could be different for houses of these types, and it is important to see if their inclusion has had a noticeable effect on the fitted regression. A plot of Cook's distances versus diagonal elements of the hat matrix (Figure 3.4) shows that not all of these leverage points would change the estimated coefficients very much if they were omitted, although as noted above change in estimated coefficients is not the only possible effect. One of those effects, in fact, is on the model selection process discussed in Section 2.3.1; once observations are omitted this is a new data set, and the "best" model might not be the same as it was before, so best subsets regression needs to be rerun.

In fact, the best three-predictor model is still based on the number of bathrooms, the living area, and the year the house was built, but there is a

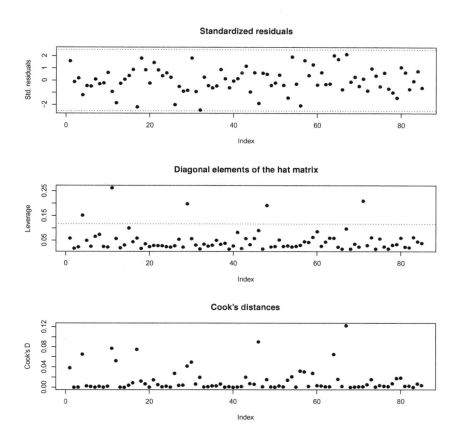

FIGURE 3.3 Index plots of diagnostics for the three-predictor regression fit for the home prices data given on page 35, with the guideline values superimposed on the standardized residuals and leverage plots.

noticeable change from the previous output (given on page 35), in that the evidence that the year the house was built adds to the predictive power of the model is noticeably weaker (because the standard error of the coefficient for that variable is 50% larger than that from the model based on all of the data), and the coefficient for living area is roughly 1.5 standard errors larger than that from the model based on all of the data:

```
Coefficients:
              Estimate Std. Error t value Pr(>|t|)    VIF
(Intercept) -9.545e+06  5.690e+06  -1.678  0.09754          .
Bathrooms    4.409e+04  1.405e+04   3.138  0.00242 1.600 **
Living.area  8.161e+01  1.790e+01   4.559 1.93e-05 1.352 ***
Year.built   4.964e+03  2.921e+03   1.699  0.09332 1.257 .
---
```

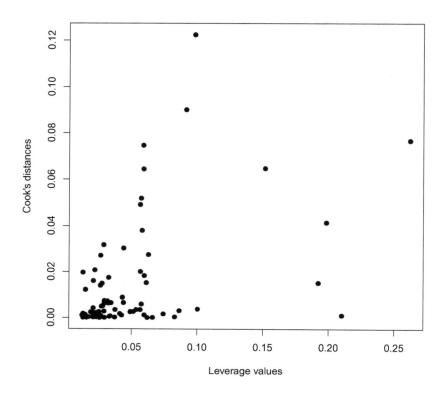

FIGURE 3.4 Plot of Cook's distance versus diagonal element of the hat matrix for the three-predictor regression fit for the home prices data given on page 35.

```
Signif. codes:
  0 '***' 0.001 '**' 0.01 '*' 0.05 '.' 0.1 ' ' 1

Residual standard error: 47390 on 76 degrees of freedom
Multiple R-squared: 0.4888,     Adjusted R-squared: 0.4687
F-statistic: 24.23 on 3 and 76 DF,  p-value: 4.192e-11
```

The apparent strength of the fit after omitting the leverage points is less (smaller R^2, larger $\hat{\sigma}$), but this is not unusual when omitting leverage points, and is certainly not a reason to not prefer this model. A viable alternative is the best two-predictor model, which removes year built as a predictor:

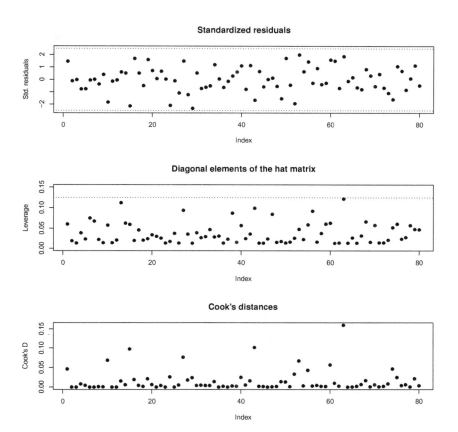

FIGURE 3.5 Index plots of diagnostics for the two-predictor regression fit for the home prices data, with the guideline values superimposed on the standardized residuals and leverage plots.

```
Coefficients:
              Estimate Std. Error t value Pr(>|t|)    VIF
(Intercept) 124450.24    26978.18   4.613 1.55e-05          ***
Bathrooms    54883.41    12689.09   4.325 4.52e-05  1.274 ***
Living.area     74.29       17.59   4.224 6.51e-05  1.274 ***
---
Signif. codes:
  0 '***' 0.001 '**' 0.01 '*' 0.05 '.' 0.1 ' ' 1

Residual standard error: 47970 on 77 degrees of freedom
Multiple R-squared: 0.4694,    Adjusted R-squared: 0.4556
F-statistic: 34.06 on 2 and 77 DF,  p-value: 2.532e-11
```

The regression diagnostics are no longer flagging any points (Figure 3.5), and residual plots (not given) now look fine. The implications of this model are not very different from the earlier one, but its simpler form could be useful; for example, it might be that a model that is not based on the year the house is built could be applicable to other neighborhoods that are similar to Levittown, but where the houses were built at a different time.

3.5   Summary

The identification of unusual observations is an important part of any regression analysis. Outliers and leverage points can have a large effect on a fitted regression, on estimated coefficients, measures of the strength of the regression, and on the model building process itself. Further, unusual observations can sometimes tell the data analyst as much (or more) about the underlying random process as the other observations. They can highlight situations in which the model could be enriched to account for their different behavior.

Residual plots can often identify unusual observations, but they should be supplemented with examination of diagnostics like the diagonal element of the hat matrix (leverage) and Cook's distance. Guidelines for what constitutes an unusual value are useful, but the values should also be plotted to make sure relative (rather than only absolute) unusualness is also made apparent. Identification of multiple unusual values can be challenging because of the masking effect, and is a topic of ongoing research.

KEY TERMS

Cook's distance: A statistic that measures the change in the fitted regression coefficients when an observation is dropped from the regression analysis, relative to the inherent variability of the coefficients. This is also equivalent to the change in the predicted value of an observation when it is included or excluded in the analysis. It is used as an influence measure for a particular observation.

Hat matrix: A matrix that contains the weights of the predictor variables that determine the predicted values. The diagonal element h_{ii} represents the potential effect of the ith observation on its own fitted value, and is used as a measure of leverage.

Influential point: An observation whose deletion causes major changes in the fitted regression. Such points exercise an undue amount of influence on the fit, thereby distorting it. As is true of all unusual observations, attention should be paid to these points.

Leverage point: A point that has a high value of the diagonal of the hat matrix. Leverage points draw the fitted regression towards them, thereby potentially distorting the fitted model. They also can have a strong effect on measures of the strength of the observed regression relationship. As is true of all unusual observations, attention should be paid to these points.

Masking effect: The tendency in data sets when there are several unusual observations clustered together that attempts to identify one observation at a time fail because they hide each other.

Outlier: An observation with a large absolute value of the residual, reflecting an observation whose response value is unusual given its predictor variable values. As is true of all unusual observations, attention should be paid to these points.

Robust regression: A regression procedure that is not affected by extreme points. This is accomplished by giving less weight to high leverage points and outliers in the fitting procedure. Instead of minimizing the sum of squared residuals, other functions of the residuals are minimized that yield the desired properties.

Standardized residual: A residual that is standardized by scaling it with its estimated standard deviation. The residuals can be standardized using the estimate of the standard deviation based on the entire data set. A modification to this is to standardize the residual using the standard deviation obtained by omitting the observation for which the residual is being computed. This is called the externally studentized residual. Since the two residuals are monotonic functions of each other, residual plots using either form will have a similar appearance.

CHAPTER FOUR

Transformations and Linearizable Models

4.1 Introduction

The linear regression models discussed thus far are only appropriate in the situation where the relationship between the response and the predictors is at least roughly linear. One situation that violates this assumption can be handled easily: the possibility of polynomial relationships. For example, examination of the data might uncover a parabolic (quadratic) relationship between x and y. This suggests enriching the model to include both linear and quadratic terms; that is, just fit a model that includes the two predictors x and x^2. This is not a problem, as this quadratic relationship with x just corresponds to a linear multiple regression relationship with those two predictors. Another possibility is the situation where the context of the problem implies the existence of an inherently nonlinear relationship between the response and the predictors. This requires moving away from linear regression to nonlinear regression methods, and is the subject of Chapter 11.

Between these two extremes are what are called **linearizable** models. These correspond to response/predictor relationships that are nonlinear, but can be changed into linear relationships through the use of transformation (most often the logarithmic transformation). Happily, it is often the case that taking logs of a variable can have multiple positive effects on a regression fit. First, variables that have most of their values relatively small and positive, but have some values much larger, so the variable covers several orders of magnitude (that is, have a distribution with a long right tail) often become much more symmetrical when treated after taking logs. In particular, previously indiscernible structure can become apparent when working in the logged scale.

Another benefit of taking logs of the response variable is that it can address certain kinds of heteroscedasticity. A situation where the variability of the response variable reflects multiplicative, rather than additive errors, can be accommodated by taking logs. Consider a true relationship that has the form

$$y_i = f(\mathbf{x}_i, \boldsymbol{\beta}) \times \delta_i, \tag{4.1}$$

where δ_i is an error term with $E(\delta_i) = 1$. In this situation the standard deviation of y_i equals $f(\mathbf{x}_i, \boldsymbol{\beta})$, so observations with larger response values will also have more variability. Equation (4.1) implies that

$$\log y_i = \log[f(\mathbf{x}_i, \boldsymbol{\beta})] + \log \delta_i \equiv \log[f(\mathbf{x}_i, \boldsymbol{\beta})] + \varepsilon_i; \tag{4.2}$$

if $\log[f(\mathbf{x}_i, \boldsymbol{\beta})]$ is a roughly linear function of the x_i variables and ε_i has constant variance σ^2, this is reasonably represented with a linear regression model based on $\log y$ as the response variable. In the next two sections we describe two relationships that satisfy this property. Note that the base of the logarithm does not matter in equation (4.2); the most common choices are the common logarithm (base 10) and the natural logarithm (base e).

The fact that multiplicative relationships become additive when taking logs implies that the standard deviation σ of the errors ε_i in equation (4.2) has a multiplicative, rather than additive, interpretation. The interval $\pm 2\hat{\sigma}$ is still a rough 95% prediction interval for the response, but since that response is $\log y$ rather than y, the correct interpretation is that y can be predicted 95% of the time to within a multiplicative factor of $10^{2\hat{\sigma}}$ (i.e., multiplying or dividing the estimated expected y by $10^{2\hat{\sigma}}$) if common logs are used, or of $e^{2\hat{\sigma}}$ if natural logs are used. Note that the intervals will be identical in the y scale in either case, as the values of $\hat{\sigma}$ are automatically adjusted accordingly.

A related point to this connection between logarithms and multiplicative relationships is that money tends to operate multiplicatively rather than additively. For example, when making an investment, people understand that it is a proportional, rather than absolute, return that is sensible (that is, an investor would not expect to be told something like "This investment could yield a profit of $50,000 for you," as it would depend on how much was invested, but would expect to be told something like "This investment could yield a profit of 10% on your investment"). For this reason, it is often useful to treat money variables in the logged scale.

4.2 Concepts and Background Material: The Log-Log Model

A model often used to describe a growth process is

$$y = \alpha x^\beta.$$

This corresponds to a *multiplicative/multiplicative* relationship, in the sense that it is consistent with proportional changes in x being associated with proportional changes in y. It is apparent that for this relationship multiplying x by a constant multiplies y by a constant, as $x \to x' = ax$ implies $y \to y' = \alpha(ax)^\beta = a^\beta \alpha x^\beta = a^\beta y$. This functional form is linearizable, since if logs are taken of both sides of the equation we obtain

$$\log y = \log \alpha + \beta \log x \equiv \beta_0 + \beta_1 \log x. \tag{4.3}$$

That is, the model is linear after logging both x and y, and is hence called the **log-log model**. The log-log model has an important interpretation in terms of demand functions. Let y above represent demand for a product, and x be the price. The price elasticity of demand is defined as the proportional change in demand for a proportional change in price; that is,

$$\frac{dy/y}{dx/x} = \frac{dy/dx}{y/x},$$

where dy/dx is the derivative of y with respect to x. Some calculus shows that for the log-log model, the elasticity is (a constant) β, and the log-log model is therefore sometimes called the **constant elasticity model**, since such a slope coefficient corresponds to an elasticity. This provides a direct interpretation for the slope coefficient in a log-log model as the proportional change in y associated with a proportional change in x (holding all else in the model fixed). Thus, when fitting model (4.3) the estimated coefficient $\hat{\beta}_j$ implies that a 1% change in x_j is associated with an estimated $\hat{\beta}_j\%$ change in y (holding all else in the model fixed if there are other predictors).

4.3 Concepts and Background Material: Semilog Models

Semilog models correspond to the situation where either the response variable or a predicting variable is logged, but not both. The two situations have fundamentally different interpretations, and so are treated separately.

4.3.1 LOGGED RESPONSE VARIABLE

Another model often used to describe a growth process is

$$y = \alpha \beta^x.$$

This corresponds to an *additive/multiplicative* relationship, in the sense that it is consistent with additive changes in x being associated with multiplicative changes in y. For this relationship adding a constant to x multiplies y by a constant, as $x \to x' = x + a$ implies $y \to y' = \alpha \beta^{x+a} = \beta^a \alpha \beta^x = \beta^a y$. If the predictor is time, another interpretation of this model is through the fact that this relationship is consistent with a growth rate that is proportional to the current level of the response. This functional form is linearizable, since if logs are taken of both sides of the equation we obtain

$$\log y = \log \alpha + x \log \beta \equiv \beta_0 + \beta_1 x, \tag{4.4}$$

corresponding to a relationship where $\log y$ is linearly related to x. An equivalent representation of this relationship is

$$y = \exp(\beta_0 + \beta_1 x). \tag{4.5}$$

This model is particularly appropriate, for example, for modeling the growth of objects over time, such as the total amount of money in an investment as a function of time, or the number of people suffering from a disease as a function of time. Growth operates multiplicatively, but time operates additively. In this situation the estimated slope $\hat{\beta}_j$ is a **semielasticity**, and $10^{\hat{\beta}_j}$ (using common logs) or $e^{\hat{\beta}_j}$ (using natural logs) is interpreted as the estimated expected multiplicative change in y associated with a one unit increase in x_j holding all else in the model fixed.

4.3.2 LOGGED PREDICTOR VARIABLE

The other possibility for a semilog model is a regression model where the response variable y is not logged, but the predictor x is. The functional relationship this implies between y and x is

$$\exp(y) = \alpha x^\beta,$$

and thus corresponds to a *multiplicative/additive* relationship, with multiplicative changes in x being associated with additive changes in y. Logging both sides gives the relationship

$$y = \log \alpha + \beta \log x \equiv \beta_0 + \beta_1 \log x.$$

The slope β_j in such a model is based on the usual interpretation of regression slopes, except that adding one to $\log x$ corresponds to multiplying x by 10 (if common logs are used) or by e (if natural logs are used). That is, for example for common logs, the model implies that multiplying x_j by 10 is associated with an estimated expected increase of $\hat{\beta}_j$ in y, holding all else in the model fixed. Such a model is appropriate in the situation where the response variable does not vary over a wide range while the predictor does. For example, a model exploring the relationship for a sample of countries between a health outcome like life expectancy and a measure of wealth like gross national income is a situation where it is reasonable to think that a proportional increase in income would be associated with an additive change in life expectancy (implying that absolute increases in income are "worth more" for low-income countries than they are for high-income countries).

4.4 Example — Predicting Movie Grosses After One Week

The movie industry is a business with a high profile, and a highly variable revenue stream. In 2010, moviegoers spent more than \$10 billion at the U.S. box office alone. A single movie can be the difference between tens of millions of dollars of profits or losses for a studio in a given year. It is not surprising, therefore, that movie studios are intensely interested in predicting revenues from movies; the popular nature of the product results in great interest in gross revenues from the general public as well.

The opening weekend of a movie's release typically accounts for 35% of the total domestic box office gross, so we would expect that the opening weekend's grosses would be highly predictive for total gross. In fact, this understates the importance of the opening weekend. It is on the strength of the opening weekend of general release that many important decisions pertaining to a film's ultimate financial destiny are made. Since competition for movie screens is fierce, movie theater owners do not want to spend more than the contractually obligatory two weeks on a film that doesn't have "legs." Should a film lose its theatrical berth very quickly, chances are slim that it will have significant play internationally (if at all), and it is less likely that it will make it to pay-per-view, cable, or network television. This all but guarantees that ancillary revenue streams will dry up, making a positive return on investment virtually impossible to achieve, as ancillary deals are predicated on domestic box office gross. Exhibitors often make the decision to keep a film running based on the strength of its opening weekend. The ability to predict total domestic gross after the first weekend of release is thus of great importance.

The following analysis is based on the movies released in the U.S. during 2009 that opened on more than 500 screens, which can be viewed as a sample from the ongoing process of movie production and release. The response variable is the total domestic (U.S.) grosses, while potential predictors are opening weekend gross (in millions of dollars), the number of screens on which the movie opened, the estimated production budget when reported (in millions of dollars), and the rating of the movie at the film review aggre-

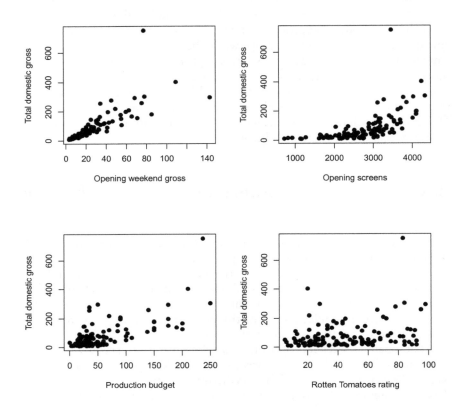

FIGURE 4.1 Scatter plots of total domestic gross versus opening weekend gross (both in millions of dollars), number of opening screens, estimated production budget (in millions of dollars), and Rotten Tomatoes rating, respectively, for 2009 movies data.

gator website Rotten Tomatoes (rottentomatoes.com). Note that the first three predictors would certainly be available to a producer after the opening weekend, and a general perception of the critical reaction to a movie would be also, even if the exact Rotten Tomatoes rating might not be. Figure 4.1 exhibits for several of the variables typical signs that the variables are better analyzed in the logged scale: plots involving total domestic gross, opening weekend gross, and budget all show bunching in the lower left corner, with a gradual spreading out of the data points moving towards the upper right corner. The relationship between total domestic gross and opening screens also looks distinctly nonlinear.

Figure 4.2 gives corresponding plots logging (base 10) the total domestic gross, first weekend gross, and budget variables. It is apparent that the relationships look much more consistent with the assumptions of linear least

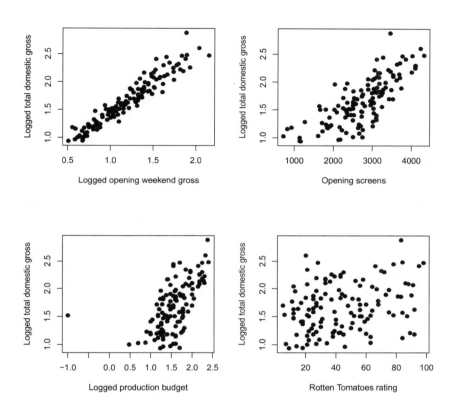

FIGURE 4.2 Scatter plots of logged total domestic gross versus logged opening weekend gross, number of opening screens, estimated logged production budget, and Rotten Tomatoes rating, respectively, for 2009 movies data. All logs are base 10.

squares regression. There are reasonably strong relationships with all of the variables other than the Rotten Tomatoes rating. There is evidence of non-constant variance in the plot of logged total gross versus logged opening weekend gross, and a very obviously unusual point to the left in the plot versus logged production budget (a potential leverage point that corresponds to "The Last House on the Left," which had a reported production budget of only roughly $100,000). The movie "Avatar" also shows up as unusually successful (and a potential outlier) at the top of several of the plots.

The following output summarizes results for a regression fit of logged total domestic gross on logged opening weekend gross, number of opening screens, logged estimated production budget, and Rotten Tomatoes rating.

```
Coefficients:
                   Estimate Std.Err.   t val Pr(>|t|) VIF
(Intercept)       2.353e-01 4.75e-02    4.95 2.5e-06       ***
Log.opening.gross 1.014e+00 4.87e-02   20.82 < 2e-16 2.93 ***
Screens           1.002e-06 2.61e-05    0.04 0.96947 3.27
Log.budget        8.657e-02 3.17e-02    2.73 0.00727 1.68 **
RT                1.619e-03 4.56e-04    3.55 0.00056 1.07 ***
---
Signif. codes:
   0 '***' 0.001 '**' 0.01 '*' 0.05 '.' 0.1 ' ' 1

Residual standard error: 0.1191 on 115 degrees of freedom
Multiple R-squared: 0.9285,    Adjusted R-squared: 0.926
F-statistic: 373.3 on 4 and 115 DF,  p-value: < 2.2e-16
```

The model fit is very strong, with R^2 over 92%, and a highly statistically significant F-statistic. While logged opening weekend gross, logged budget, and Rotten Tomatoes rating are all strongly statistically significant, given the other predictors the number of screens on which the movie opens does not add any predictive power. This is consistent with the results of a best subsets regression (output not given), which identifies the model with all of the predictors other than number of screens as best. The output for this simplified model is given below.

```
Coefficients:
                   Estimate Std.Err.   t val Pr(>|t|) VIF
(Intercept)       0.2358772 0.04474    5.27 6.3e-07       ***
Log.opening.gross 1.0154911 0.03440   29.52 < 2e-16 1.47 ***
Log.budget        0.0870212 0.02924    2.98 0.00355 1.44 **
RT                0.0016163 0.00045    3.61 0.00046 1.04 ***
---
Signif. codes:
   0 '***' 0.001 '**' 0.01 '*' 0.05 '.' 0.1 ' ' 1

Residual standard error: 0.1185 on 116 degrees of freedom
Multiple R-squared: 0.9285,    Adjusted R-squared: 0.9267
F-statistic: 502.1 on 3 and 116 DF,  p-value: < 2.2e-16
```

The estimated standard deviation of the errors $\hat{\sigma} = .1185$, but as was noted on page 38, this should be adjusted to account for the model selection process. In this case the adjustment proposed in equation (2.4) makes little difference, as $\tilde{\sigma} = .119$. This implies that these variables can predict total domestic gross to within a multiplicative factor of 1.73, roughly 95% of the time ($10^{(2)(.119)} = 1.73$). Thus, it would not be surprising for a movie predicted to have a total gross of $100 million to have an actual gross as large as $173 million or as small as $58 million, which reflects the inherent high variability in movie grosses, even based on a model that accounts for more than 92% of the variability in logged grosses. The coefficients for logged opening weekend gross and logged budget are elasticities, so they imply that holding

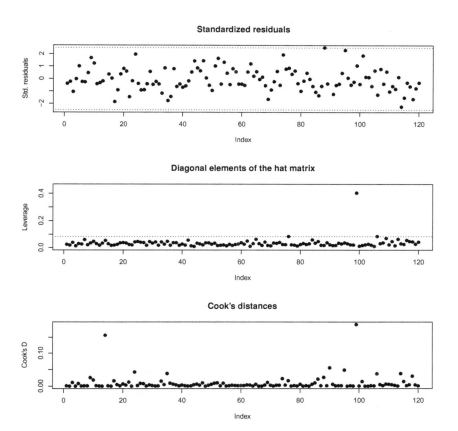

FIGURE 4.3 Index plots of diagnostics for the regression fit on the three-predictor model for the movie grosses data, with guideline values superimposed on the standardized residuals and leverage plots.

all else fixed a 1% increase in opening weekend gross is associated with an estimated expected 1.02% increase in total domestic gross, and a 1% increase in production budget is associated with an estimated expected 0.09% increase in total domestic gross, respectively. The coefficient for Rotten Tomatoes rating is a semielasticity, implying that holding all else fixed an increase of one point in the rating is associated with an estimated expected 0.4% increase in total domestic gross ($10^{.0016} = 1.004$).

Regression diagnostics (Figure 4.3) identify "The Last House on the Left" as an extreme leverage point, as would be expected. After omitting this point, best subsets regression still points to the same three-predictor model as best, and the regression results change very little, although diagnostics (Figure 4.4) now indicate several marginal leverage points (corresponding to "Next Day Air," "Ponyo," and "The Twilight Saga: New Moon").

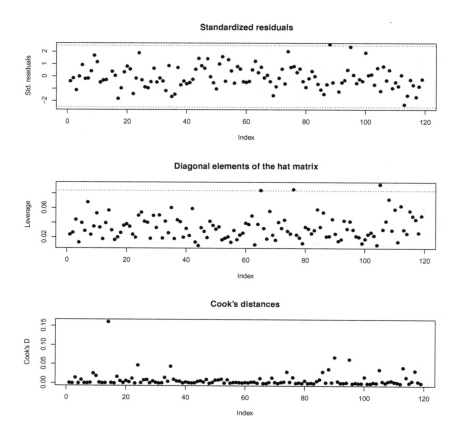

FIGURE 4.4 Index plots of diagnostics for the three-predictor regression fit after removing the leverage point, with guideline values superimposed on the standardized residuals and leverage plots.

```
Coefficients:
                    Estimate Std.Err.   t val Pr(>|t|) VIF
(Intercept)        0.2156550  0.04862   4.44 2.1e-05       ***
Log.opening.gross  1.0006562  0.03712  26.96 < 2e-16 1.72 ***
Log.budget         0.1120680  0.03759   2.98 0.00350 1.71 **
RT                 0.0015644  0.00045   3.47 0.00073 1.05 ***
---
Signif. codes:
  0 '***' 0.001 '**' 0.01 '*' 0.05 '.' 0.1 ' ' 1

Residual standard error: 0.1185 on 115 degrees of freedom
Multiple R-squared: 0.9291,    Adjusted R-squared: 0.9273
F-statistic: 502.4 on 3 and 115 DF,  p-value: < 2.2e-16
```

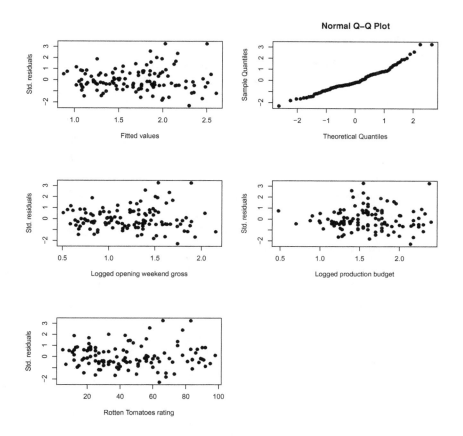

FIGURE 4.5 Residual plots for the three-predictor regression fit after removing the leverage point for the movie grosses data.

Residual plots (Figure 4.5) illustrate more serious issues: residuals that are somewhat right-tailed, and nonconstant variance, where movies with larger estimated (logged) domestic grosses have higher variability. The latter issue requires the use of **weighted least squares**, which will be discussed in Sections 6.3.3 and 10.7.

TABLE 4.1 Total domestic grosses (in millions of dollars) for 2010 movies with 95% prediction limits, based on using the model fit to 2009 data.

Movie	Lower limit	Actual gross	Upper limit
Cop Out	27.88	44.88	83.03
Daybreakers	25.70	30.10	76.70
Dear John	46.17	80.01	138.26
Edge of Darkness	31.73	43.31	94.15
Extraordinary Measures	9.27	12.48	27.69
From Paris With Love	13.82	24.08	41.27
Leap Year	13.16	25.92	39.21
Legion	25.84	40.17	77.01
Percy Jackson and the Olympians	59.27	88.77	176.32
Shutter Island	82.05	128.01	244.73
The Book of Eli	60.85	94.84	180.81
The Crazies	27.56	39.12	82.50
The Tooth Fairy	21.82	60.02	65.12
The Wolf Man	59.30	62.19	178.00
Valentine's Day	88.63	110.49	267.54
Youth in Revolt	11.57	15.29	34.63

A natural use of this model is to forecast grosses for future movies. Table 4.1 summarizes the results of such predictions based on this model for movies released in January and February of 2010, giving 95% prediction limits for total gross (obtained by antilogging the upper and lower limits of prediction intervals based on the model for logged total domestic gross). It can be seen that the model does a good job of predicting future grosses, with all of the prediction intervals containing the actual total domestic gross values.

4.5　Summary

Linear least squares modeling makes several assumptions about the underlying regression relationship in the population, which might not hold. Several violations of these assumptions, including multiplicative rather than additive errors (which result in nonconstant variance) and certain forms of nonlinearity, correspond to linearizable violations, in that using the target or predicting variables (or both) in the logged scale can address them. The logged transformation is particularly useful in situations where a variable is long right-tailed.

When the target variable is logged the slope coefficients take the form of elasticities (for logged predictors) or semielasticities (for unlogged predictors) that have natural and intuitive interpretations in the original scale, making this transformation a particularly attractive one in many situations.

KEY TERMS

Elasticity: The proportional change in the response corresponding to a proportional change in a predictor. In economics this often corresponds to a proportional change in demand of a product or service corresponding to a proportional change in its price.

Log-log model: A relationship of the form $E(y) = \alpha x^\beta$. This relationship is linearizable, as the unknown parameters can be estimated from a linear regression of $\log y$ on $\log x$.

Semielasticity: The proportional change in the response corresponding to an additive change in a predictor.

Semilog models: Relationships of the form $y = \alpha\beta^x$ or $\exp(y) = \alpha x^\beta$. These relationships are linearizable, as the unknown parameters can be estimated from a linear regression of $\log y$ on x or y on $\log x$, respectively.

CHAPTER FIVE

Time Series Data and Autocorrelation

5.1   Introduction

As was noted in Section 1.2.3, a standard assumption in regression modeling is that the random errors ε_i are uncorrelated with each other. Correlation in the residuals represents structure in the data that has not been taken into account. When the observations have a natural sequential order, and as a result the correlation structure is related to that order, this correlation is called **autocorrelation**. Although the issues we will discuss in this chapter can arise in any situation where there is a natural sequential ordering to the data, we will refer to it generically as time series data, since data ordered in time is certainly the most common application area.

Autocorrelation occurs for several reasons. In time series data it is often the case that adjacent values are similar, with high values following high values and low values following low values. Often this is a result of the target variable being subjected to similar external conditions. Adjacent errors in economic data, which correspond to measurements from consecutive time periods like days, months, or years, are often positively correlated because of the effects of underlying economic processes that are evolving over time. Adjacent experimental values, such as successive outputs in a production process, can be positively correlated because they are affected by similar short-term conditions on machinery.

Autocorrelation in time series data also can arise by omission of an important predictor variable from the model. If the successive values of an important omitted variable are correlated, the errors from a model that omits this variable will tend to be autocorrelated, since the errors will reflect the effects of the missing variable. This means that issues of model selection for time series data can become conflated to some extent with issues of autocorrelation. Measures designed to compare models that do not account for autocorrelation can be misleading in the presence of autocorrelation, complicating the ability to identify appropriate choices of variables.

In this chapter we will discuss some of the issues related to building regression models for time series data. We will first discuss the effects of autocorrelation if it is ignored. We will then examine several approaches to identifying autocorrelation, which range from one requiring strong assumptions (the Durbin-Watson statistic) to one related to a simple graphical examination of residuals that requires virtually no assumptions (the runs test). We then discuss several relatively simple approaches to accounting for common forms of autocorrelation, including trends and seasonal effects, and explore how values from previous time periods can be used to enrich a regression model and account for autocorrelation. We conclude with discussion of a more sophisticated approach to handling autocorrelation that moves past ordinary least squares estimation to estimation designed for time series data.

It is important to note that the methods discussed here only scratch the surface of time series analysis and modeling. It would not be at all surprising if an analyst finds that using the methods discussed in this chapter will account for much (or even most) of the observed autocorrelation in a data set, but in many cases there will still be apparent autocorrelation that requires more complex methods. That is beyond the scope of this book, but there are many books that discuss such methods that can be consulted, such as Cryer and Chan (2008) and Kedem and Fokianos (2002).

5.2   Concepts and Background Material

Autocorrelation can have several problematic effects on a fitted model. These are as follows:

1. Least squares estimates are unbiased, but are not efficient, in the sense that they do not have minimum variance. Since the degree to which the OLS estimates are inefficient depends on the type and amount of autocorrelation in both the errors and the predictor, we need to specify a particular form of autocorrelation to explore this. A **first-order autoregressive** [AR(1)] process satisfies

$$\varepsilon_i = \rho\varepsilon_{i-1} + z_i, \quad |\rho| < 1, \tag{5.1}$$

where the z_i are independent and identically distributed normally distributed random variables. The standard assumption underlying least squares is that $\rho = 0$, and in that situation the OLS estimator has the minimum variance possible among all unbiased estimators, but if $\rho \neq 0$ that is no longer the case [in this case $\rho = \mathrm{corr}(\varepsilon_i, \varepsilon_{i-1})$]. Figure 5.1 illustrates this point graphically for a simple regression model where the predictor x follows an AR(1) process with parameter λ. The panels of the plot correspond to values of λ equal to 0, 0.1, 0.3, 0.5, 0.7, and 0.9 from the lower left to the upper right, with each plot giving the ratio of the variance of the OLS estimate of β_1 to that of the unbiased estimator with minimum variance for large samples versus the autocorrelation of the errors ρ. It can be seen that while the OLS estimator is not very inefficient for all λ when $|\rho| < 0.3$ (with plotted inefficiency close to the horizontal line corresponding to a ratio equal to 1), for larger amounts of autocorrelation in the errors its variance can be considerably larger than the minimum possible value. In Section 5.4.5 we will discuss construction of the estimator that has minimum variance for all ρ for this situation.

2. The estimates of the standard errors of the regression coefficients and of σ^2 are biased. Figure 5.2 illustrates this point. Each panel of the plot gives the percentage bias of the usual estimate of the variance of the OLS $\hat{\beta}_1$ for large samples as a function of ρ for a given λ. The solid horizontal line at 0 corresponds to no bias, while the dashed line corresponds to a bias of -100% (the lowest possible value). It can be seen that when ρ and λ are nonzero with the same sign (the right side of each panel) the estimated variance is negatively biased (often extremely biased). This means that t-statistics will be larger than they should be (since the square root of the estimated variance is the denominator of the t-statistic), resulting in a spurious impression of precision. This is the typical situation for economic data (where both ρ and λ will be positive). If the two autocorrelation parameters are of opposite sign (the left side of each panel) the estimated variances are positively biased, and measures of the strength of the regression will be too small rather than too large. Note that while there is no

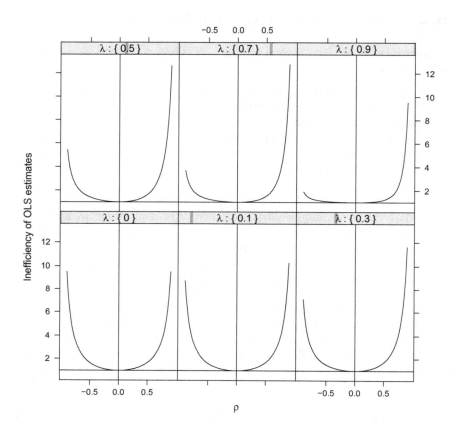

FIGURE 5.1 Inefficiency of OLS estimator compared to minimum variance estimator under first-order autoregressive errors, with first-order autoregressive predictor; $\rho = \text{corr}(\varepsilon_i, \varepsilon_{i-1})$ and $\lambda = \text{corr}(x_i, x_{i-1})$.

requirement that predictors lack autocorrelation to justify least squares (which is reflected in the bias equaling 0 when $\rho = 0$ for all values of λ), autocorrelation of the predictors does have a strong effect on the consequences of autocorrelation in the errors if it exists, with stronger effects as $|\lambda|$ increases. Not only does the direction of the bias depend on the sign of λ, the bottom left panel of the figure shows that for this type of time series structure there is no bias in the estimated variances if $\lambda = 0$ for all values of ρ.

3. As a result of this bias, confidence intervals, significance tests, and prediction intervals are no longer valid.

Given the seriousness of the autocorrelation problem, corrective action to address it should be taken. The appropriate action depends on the source

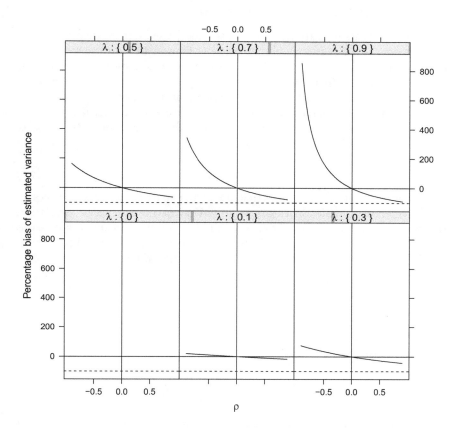

FIGURE 5.2 Percentage bias of the estimated variance of the OLS estimator of β_1 under first-order autoregressive errors, with first-order autoregressive predictor.

of the autocorrelation. If autocorrelation is due to the absence of a variable or other structure (such as seasonal effects) in the model, once the necessary structure is included, the problem is reduced or disappears. On the other hand, if autocorrelation is an inherent part of the population error structure, addressing this autocorrelation requires transformation or changing the least squares criterion.

5.3　Methodology: Identifying Autocorrelation

We describe three tests for detecting autocorrelation that range from strongly parametric (based on a set of strong assumptions about the underlying error process) to nonparametric (based on very weak assumptions).

5.3.1 THE DURBIN-WATSON STATISTIC

The **Durbin-Watson statistic** is the most widely used test for detecting auto-correlation. The test is strongly parametric, being based on the assumption that the errors follow the AR(1) process (5.1). The situation where there is no autocorrelation ($\rho = 0$) corresponds to all of the standard least squares assumptions: that is, the errors are independent and identically distributed Gaussian random variables. Further, if autocorrelation exists, it is assumed that it takes the specific AR(1) form. It can be shown that the AR(1) process on the errors implies that the autocorrelations geometrically decay as the lag increases; that is,

$$\rho_k \equiv \text{corr}(\varepsilon_i, \varepsilon_{i-k}) = \rho^k. \tag{5.2}$$

While the implication that errors that are farther apart are more weakly corre-lated is often reasonable, the errors will often have a more complex correlation structure than this, so the adequacy of the Durbin-Watson statistic depends on (5.1) [and hence (5.2)] being a reasonable approximation to reality.

In its standard form the Durbin-Watson statistic d tests the hypotheses

$$H_0 : \rho = 0$$

versus

$$H_a : \rho > 0.$$

The test statistic is defined as

$$d = \frac{\sum_{i=2}^{n}(e_i - e_{i-1})^2}{\sum_{i=1}^{n} e_i^2},$$

where e_i is the ith least squares residual. A drawback of this statistic is that it is not pivotal; that is, its distribution (and hence critical values used to apply it as a hypothesis test) depends on unknown parameters. Durbin and Watson (1951) showed that the statistic is asymptotically pivotal, and further showed that a set of two critical values $\{d_L, d_U\}$ can be used to implement the test as follows:

1. If $d < d_L$, reject H_0.

2. If $d > d_U$, do not reject H_0.

3. If $d \in (d_L, d_U)$, the test is inconclusive.

Critical values $\{d_L, d_U\}$ have been tabulated by many authors and are available on the internet, including at the website for this book. Alternatively, for large n (say $n > 100$) a normal approximation to the distribution of d can be used,

$$z = \left(\frac{d}{2} - 1\right)\sqrt{n}.$$

This form of the test shows that a value of d close to 2 is indicative of the absence of autocorrelation. Some statistical packages provide an exact tail probability for the test based on its null distribution (a linear combination of χ^2 variables).

Tests for negative autocorrelation are performed more rarely than for positive autocorrelation. If a test is desired for $\rho < 0$, the appropriate test statistic is $4 - d$, and the procedure outlined above is then followed.

5.3.2 THE AUTOCORRELATION FUNCTION (ACF)

Use of the Durbin-Watson statistic is based on assuming an AR(1) process for the errors, so it is important to check whether that holds. This can be done through use of an **autocorrelation function (ACF) plot**. In this plot estimates of autocorrelations for a range of lags are plotted, often with approximate 95% confidence bands around 0 superimposed. The estimate of the kth-lag autocorrelation is

$$\hat{\rho}_k = \frac{\sum_{i=k+1}^{n} e_i e_{i-k}}{\sum_{i=1}^{n} e_i^2}.$$

If there is no autocorrelation the standard error of $\hat{\rho}_k$ satisfies $s.e.(\hat{\rho}_k) \approx 1/\sqrt{n}$, implying that $\hat{\rho}_k$ values greater than roughly $2/\sqrt{n}$ are significantly different from 0 at a .05 level. Examination of the ACF plot can show if there is any evidence of autocorrelation in the residuals, if observed autocorrelation is consistent with an AR(1) process (by seeing if the estimated autocorrelations follow a roughly geometric decay), or if other forms of autocorrelation (such as seasonality) are present.

5.3.3 RESIDUAL PLOTS AND THE RUNS TEST

The presence of autocorrelation in a given set of time series data can be detected by an examination of an index plot of the values in time order. As was noted on page 15, a standard approach for the detection of autocorrelation of regression errors is the corresponding index plot of the (standardized) residuals. This is particularly helpful in the presence of underlying positive autocorrelation of the errors, as this corresponds to a positive (negative) error in one time period being associated with a positive (negative) error in the next time period. In this case the residual plot has a distinctive cyclical pattern, where residuals of the same sign are clustered, with positive residuals tending to follow positive residuals and negative residuals tending to follow negative residuals. The corresponding pattern for negative autocorrelation, with positive residuals tending to follow negative residuals (and vice versa), also can occur, but this is difficult to see in a residual plot.

These patterns of same-signed residuals either following or not following each other is the principle underlying a nonparametric test of autocorrelation. The **runs test** formalizes the detection of residual clustering by counting the number of runs of residuals of the same sign. For example, in a series of residuals with signs $+++--+-+--------+++++-+------++$, there are $n_+ = 14$ positive residuals, $n_- = 16$ negative residuals, and $u = 9$ runs (a run of 4 positive residuals, followed by a run of 4 negative residuals, and so on). For small sample sizes the null distribution of the number of runs can be determined exactly on the basis of all possible permutations of pluses and minuses, while for larger sample sizes if there is no autocorrelation u is roughly normally distributed with mean

$$\mu = \frac{2n_+ n_-}{n} + 1$$

and variance

$$\sigma^2 = \frac{2n_+ n_- (2n_+ n_- - n)}{n^2(n-1)}.$$

Too few runs corresponds to positive autocorrelation, while too many runs corresponds to negative autocorrelation, thus providing a test of the null hypothesis that the errors are independent and identically distributed. The runs test has the advantage of being a nonparametric test, not requiring any assumptions about the underlying distribution of the errors (other than that there is a common distribution for all errors).

5.4   Methodology: Addressing Autocorrelation

5.4.1 DETRENDING AND DESEASONALIZING

The time series literature is very extensive, and it is beyond the scope of this book (and most routine regression analyses) to cover that material in detail. Fortunately, it turns out that many autocorrelation problems can be addressed to a large (if not complete) extent using relatively simple methods that should be part of the data analyst's toolkit. Indeed, simply including appropriate predictors in a regression model can often take initially strong autocorrelation in the target variable and turn it into little or no autocorrelation in the residuals, accounting for the problem almost completely.

Temporal data (observations taken over time) often have two characteristics, trend and seasonality. Trend is the general movement in the data, going up or down over the period of observation. Since many variables naturally grow over time (for example, family income because of inflation or national production because of population growth), it is reasonable to incorporate such growth into the model. This can be done in two simple ways. First, variables (especially response variables) should be modeled in a natural scale that is comparable across time periods if at all possible. Thus, if incomes grow naturally because of inflation, they should be corrected for inflation by using an appropriate price deflator so they are in constant dollars rather than current dollars. Similarly, if production grows naturally with population, population-corrected per capita measures should be used.

A second approach (which can still be useful even if the variables have been rescaled as described above) is to incorporate a time trend into the model.

The trend is often modeled by incorporating a linear or quadratic term (using time and perhaps time^2 as predictors). The need for such detrending can be assessed using the model building methods described in Chapter 2. In situations where the growth is exponential, the semilog model described in Section 4.3 is appropriate, with the target variable logged and time entering the model unlogged.

Temporal data also often have a component that varies with time, representing the effects of the season. For example, monthly sales data often exhibit the recurring pattern of sales being higher than normal at the end of the year (because of the Christmas season) and lower than normal at the beginning of the year (the aftermath of the Christmas season). This sort of pattern would show up as persistent monthly patterns in monthly data, or persistent quarterly patterns for quarterly data. It can often be identified by examining the residuals appropriately. For example, a seasonal effect in quarterly data can show up in an ACF plot as a significant autocorrelation at lag 4 (and multiples of 4), since residuals that are four quarters apart follow the persistent seasonal pattern by being one year apart; similarly, a monthly seasonal effect can show up as a significant autocorrelation at lag 12 (and multiples of 12). Side-by-side boxplots of residuals separated by quarter or month also can uncover seasonal effects.

Seasonal effects can be taken into account using a set of indicator variables as a generalization of the analysis discussed in the presence of data with two subgroups in Section 2.4 (since quarterly data fall into four distinct subgroups, monthly data fall into 12 distinct subgroups, and so on). An indicator variable is defined for each subgroup (quarter or month) with one of the indicators omitted to account for the presence of an intercept term in the model (this approach is discussed in much more detail in Chapter 6). This procedure accounts for systematic shifts in the target variable from seasonal effects (that is, it corresponds to fitting a constant shift model, with the regression (hyper)plane shifted up or down by season).

It is often the case that trend and seasonal effects are viewed as nuisance effects, not being related to the contextual relationships of interest to the researcher. Removing trend and seasonality from the data in these ways thus allows a more focused examination of the structure of the data after these effects have been taken into account.

5.4.2 EXAMPLE — E-COMMERCE RETAIL SALES

Electronic (e-)commerce is a multibillion dollar business, encompassing online sales sites (such as Amazon) and auction sites (such as eBay); indeed, it is hard to imagine almost any retail business not having some sort of online sales presence, if not now then in the near future. The importance of this business sector makes it important to understand the dynamics of e-commerce sales. Figure 5.3 gives a time series plot of the quarterly e-commerce retail sales (in millions of dollars) for the United States from the fourth quarter of 1999 through the first quarter of 2011, based on information from the U.S. Cen-

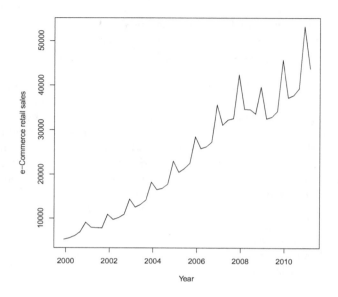

FIGURE 5.3　Time series plot of quarterly U.S. e-commerce retail sales.

sus Bureau. Two characteristics of this series are immediately apparent: sales trended upwards during the decade, and there is a clear seasonal effect, with sales peaking in the fourth quarter and then dropping sharply in the first quarter (this of course corresponds to the effects of holiday shopping).

These patterns are characteristic of retail sales in general, so it is reasonable to think that total sales could be a good predictor for e-commerce sales (this assumes, of course, that the underlying relationships during this time period remain at least roughly the same in the future). Figure 5.4 is a scatter plot of e-commerce sales versus total sales, and it is clear that there is the expected direct relationship between the two.

Output for the regression of e-commerce sales on total sales is given below.

```
Coefficients:
              Estimate Std. Error  t value  Pr(>|t|)
(Intercept) -7.396e+04  8.120e+03   -9.108  1.10e-11 ***
Total.sales  1.103e-01  9.109e-03   12.104  1.35e-15 ***
---
```

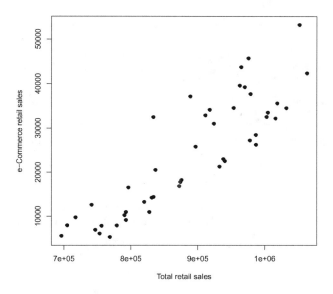

FIGURE 5.4 Scatter plot of e-commerce retail sales versus total retail sales.

```
Signif. codes:
  0 '***' 0.001 '**' 0.01 '*' 0.05 '.' 0.1 ' ' 1

Residual standard error: 6270 on 44 degrees of freedom
Multiple R-squared: 0.769,      Adjusted R-squared: 0.7638
F-statistic: 146.5 on 1 and 44 DF,  p-value: 1.348e-15
```

There is a highly statistically significant relationship between e-commerce sales and total sales. Unfortunately, autocorrelation is apparent in the residuals, as Figure 5.5 shows. The time series plot of the standardized residuals in plot (a) shows a cyclical pattern indicative of positive autocorrelation. The ACF plot in plot (b) shows that there are significant autocorrelations at many lags, including lags 1–4 and 8.

The jumps in the autocorrelation at lags 4 and 8 are particularly interesting, since the quarterly nature of the data implies that these are likely to represent seasonality. This is further supported in Figure 5.6, which is a set of side-by-side boxplots of the standardized residuals separated by quarter. Remarkably, however, the seasonal effect is **not** that sales are higher than expected in the fourth quarter and lower than expected in the first quarter, as the original time series plot of e-commerce sales would have implied; rather, e-commerce sales are **higher** than expected in the first quarter. The reason for this is that while the holiday seasonal effect does exist in the e-commerce sales,

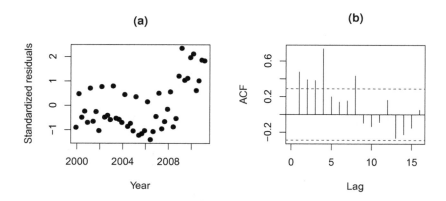

FIGURE 5.5 Residual plots for the e-commerce regression fit based on total retail sales. (a) Time series plot of standardized residuals. (b) ACF plot of residuals, with superimposed 95% confidence limits around 0.

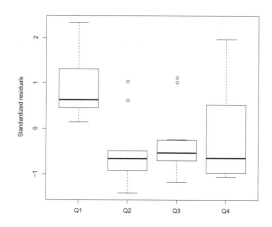

FIGURE 5.6 Side-by-side boxplots of the standardized residuals for the e-commerce regression fit based on total retail sales.

it is actually weaker than the corresponding effect in the total sales; when total sales are taken into account, first quarter e-commerce sales look better than expected because they are not as relatively poor as sales overall. This highlights that the autocorrelation structure in the residuals, which take predictor(s) into account, can be very different from that in the original response variable.

This seasonality effect can potentially be accounted for by adding indicator variables for any of the three quarters. The resultant output (based on including indicators for the first three quarters) is given below.

```
Coefficients:
              Estimate  Std. Error  t value  Pr(>|t|)
(Intercept)  -8.538e+04  7.667e+03  -11.136  5.68e-14  ***
Total.sales   1.217e-01  8.129e-03   14.972  < 2e-16   ***
Quarter.1     7.564e+03  2.275e+03    3.325  0.00187   **
Quarter.2    -2.108e+03  2.199e+03   -0.959  0.34328
Quarter.3    -7.620e+02  2.203e+03   -0.346  0.73122
---
Signif. codes:
   0 '***' 0.001 '**' 0.01 '*' 0.05 '.' 0.1 ' ' 1

Residual standard error: 5241 on 41 degrees of freedom
Multiple R-squared: 0.8496,    Adjusted R-squared: 0.835
F-statistic: 57.92 on 4 and 41 DF,  p-value: 2.49e-16
```

The quarterly indicators add significant predictive power (the partial F-test comparing the models with and without the seasonal indicators is $F = 7.33$ on $(3, 41)$ degrees of freedom, with $p = .0005$, reflecting the increase in R^2 from 77% to 85%). The coefficient for Quarter.1 implies that first quarter e-commerce sales are estimated to be on average $7.56 billion higher than fourth quarter sales, given total sales are held fixed.

Unfortunately, while this has addressed the seasonality to a large extent, it has not addressed all of the problems. Figure 5.7 shows that there is a clear break in the residuals, with e-commerce sales higher than expected starting with the fourth quarter of 2008. This is reflecting the relative insensitivity of e-commerce sales to the worldwide recession that began in late 2008; while total retail sales went down during that time, e-commerce sales continued to go up. It is also interesting to note that the ACF plot [Figure 5.7(b)] is unable to identify this pattern, instead pointing to an AR(1)-like slow decay of the autocorrelations. This reinforces the importance of looking at the residuals in different graphical ways, and not just depending on test statistics to identify problems.

The following output summarizes a regression model that adds a constant shift corresponding to quarters during the recession. The indicator for the recession is highly statistically significant, implying $11.9 billion higher estimated expected e-commerce sales during the recession given the quarter and the total retail sales.

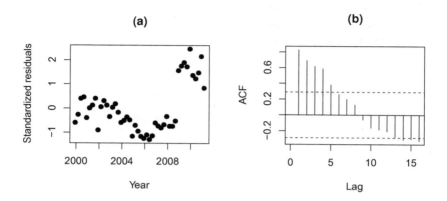

FIGURE 5.7 Residual plots for the e-commerce regression fit based on total retail sales and quarterly indicators. (a) Time series plot of standardized residuals. (b) ACF plot of residuals.

```
Coefficients:
                 Estimate  Std. Error  t value  Pr(>|t|)
(Intercept)    -7.208e+04   2.937e+03  -24.540  < 2e-16  ***
Total.sales     1.041e-01   3.181e-03   32.721  < 2e-16  ***
Quarter.1       5.888e+03   8.432e+02    6.983  1.98e-08 ***
Quarter.2      -1.773e+03   8.090e+02   -2.191   0.0343  *
Quarter.3      -5.199e+02   8.106e+02   -0.641   0.5250
Recession       1.193e+04   7.356e+02   16.217  < 2e-16  ***
---
Signif. codes:
  0 '***' 0.001 '**' 0.01 '*' 0.05 '.' 0.1 ' ' 1

Residual standard error: 1928 on 40 degrees of freedom
Multiple R-squared: 0.9802,     Adjusted R-squared: 0.9777
F-statistic:   395 on 5 and 40 DF,  p-value: < 2.2e-16
```

At this point the R^2 of the model is over 98%, and most of the autocorrelation is accounted for (see Figure 5.8). There are three quarters with unusually high e-commerce sales (the fourth quarters of 2007, 2009, and 2010, respectively), but otherwise the plot of standardized residuals versus fitted values, time series plot of the standardized residuals, and normal plot of the standardized residuals look reasonable. The ACF plot of the residuals flags a significant autocorrelation at lag 4, suggesting some sort of seasonality that was not accounted for by the indicators. This is quite possible, as much more complex deseasonalizing methods (such as the X-12 ARIMA deseasonalizing method used by the U.S. Census Bureau; U.S. Census Bureau, 2011) are often used with economic data. Side-by-side boxplots of the standardized residuals suggest potential heteroscedasticity related to both quarter (with fourth quar-

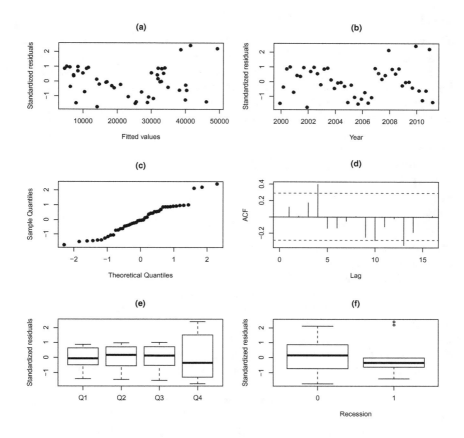

FIGURE 5.8 Residual plots for the e-commerce regression fit based on total retail sales, quarterly indicators, and recession indicator. (a) Plot of standardized residuals versus fitted values. (b) Time series plot of standardized residuals. (c) Normal plot of standardized residuals. (d) ACF plot of residuals. (e) Side-by-side boxplots of standardized residuals separated by quarter. (f) Side-by-side boxplots of standardized residuals separated by pre- or post-2008 recession.

ter residuals more variable) and recession (with pre-recession quarters more variable); ways to address such heteroscedasticity is the subject of Section 6.3.3.

Given the apparent non-AR(1) nature of the autocorrelation and heteroscedasticity the Durbin-Watson statistic is not appropriate. The runs test, however, is (virtually) always appropriate, and in this case when applying it to the residuals it has $p = .008$, reinforcing that there is still some autocorrelation present in the residuals.

5.4.3 LAGGING AND DIFFERENCING

The values of variables from previous time periods can often be useful in modeling the value from a current time period in time series data. This can reflect a natural time lag of the effect of a predictor on the response variable; for example, the Federal Reserve Bank changes interest rates in order to affect the dynamics of the U.S. economy, but it is expected that such effects would take one or two quarters to be felt through the economy. Thus, it would be natural to use a lagged value of interest rate as a predictor of gross national income (that is, using x_{i-1} or x_{i-2} to model y_i for quarterly data).

A different use of a lagged variable as a predictor is using a lagged version of the response variable itself as a predictor; that is, using for example y_{i-1} as a predictor of y_i, resulting in

$$y_i = \beta_0 + \beta_1 x_{1i} + \cdots + \beta_p x_{pi} + \beta_{p+1} y_{i-1} + \varepsilon_i \qquad (5.3)$$

(in principle further lags, such as y_{i-2} or y_{i-3}, could be used). Including a lagged response as a predictor will often reduce autocorrelation in the errors dramatically, as it directly models the tendency for time series values to move in a cyclical pattern. This fundamentally changes the interpretation of other regression coefficients, as they now represent the expected change in the response corresponding to a one unit change in the predictor holding the previous time period's value of the response fixed (as well as holding everything else in the model fixed), but from a predictive point of view can dramatically improve the predictive power of a model while reducing the effects of auto-correlation.

A related operation is **differencing** variables; that is, modeling changes in the response value, rather than the value itself. Formally this corresponds to a special case of (5.3) with $\beta_{p+1} = 1$. Differencing a variable also can be meaningful contextually, as in many situations the change in the level of a variable is more meaningful than the level itself. The example that follows based on stock prices is such an example, since it is returns (the proportional change in prices) that are meaningful to an investor, not prices themselves. It is standard practice in time series modeling in general to difference nonstationary series (time series where the distribution of the series changes over time) for exactly this reason.

The Durbin-Watson statistic is not meaningful for a model using lagged response values as a predictor, and should not be used in that situation.

5.4.4 EXAMPLE — STOCK INDEXES

The Standard & Poor's (S&P) stock indexes are well-known value-weighted indexes of stock prices of publicly-held firms. The S&P 500 is based on 500 large capitalization firms, the S&P 400 is based on 400 mid-capitalization firms, and the S&P 600 is based on 600 small capitalization firms. It would be expected that such indexes would move together based on the overall health of the economy, but it is not clear exactly what those relationships might be.

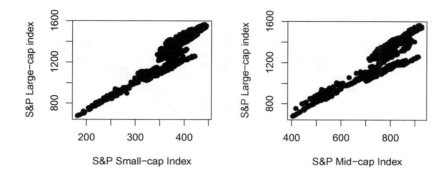

FIGURE 5.9 Scatter plots of daily S&P 500 (large-cap) index values from 2006 through 2010 versus (a) S&P 600 (small-cap) index values, and (b) S&P 400 (mid-cap) index values.

Figure 5.9 gives scatter plots of the daily S&P 500 (large-cap) index versus the S&P 600 (small-cap) and S&P 400 (mid-cap) indexes, respectively, for all trading days from 2006 through 2010. Clearly there is a direct relationship between the indexes, but the plots seem to suggest several separate regimes.

As was noted in the previous chapter, it is often the case that money data are better analyzed in the logged scale. Figure 5.10 gives corresponding scatter plots based on (natural) logged index values, which still have an apparent pattern of different regimes in the series but appear to reflect more linear relationships with less heteroscedasticity.

Regression output for the regression of logged large-cap index on logged small-cap and logged mid-cap indexes is given below.

```
Coefficients:
                   Estimate Std. Error t value Pr(>|t|)
(Intercept)         1.57402    0.05949  26.458  < 2e-16 ***
log.S.P.Small.cap   1.25458    0.03835  32.715  < 2e-16 ***
log.S.P.Mid.cap    -0.27745    0.03910  -7.096 2.14e-12 ***
---
Signif. codes:
   0 '***' 0.001 '**' 0.01 '*' 0.05 '.' 0.1 ' ' 1

Residual standard error: 0.04833 on 1256 degrees of freedom
Multiple R-squared: 0.9284,    Adjusted R-squared: 0.9283
F-statistic:  8145 on 2 and 1256 DF,  p-value: < 2.2e-16
```

The regression relationship is very strong, but autocorrelation is a serious problem. Figure 5.11 illustrates the strong cyclical pattern in the standardized residuals, consistent with a nonstationary time series that has mean value that shifts up and down. The ACF plot is consistent with this, in that the estimated

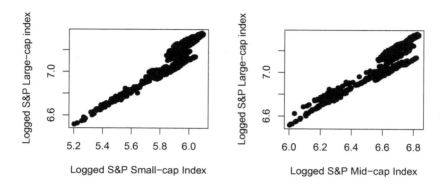

FIGURE 5.10 Scatter plots of logged daily S&P 500 (large-cap) index values versus (a) logged S&P 600 (small-cap) index values, and (b) logged S&P 400 (mid-cap) index values.

(a)   **(b)**

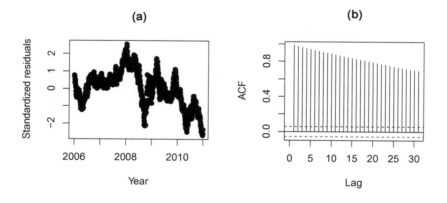

FIGURE 5.11 Residual plots for logged large-cap index regression fit based on logged small-cap and logged mid-cap indexes. (a) Time series plot of standardized residuals. (b) ACF plot of residuals.

autocorrelations of the residuals are very large, and decay very slowly. The Durbin-Watson statistics is $d = .026$, which is extremely strongly statistically significant.

Both context and the statistical results imply that differencing the data is appropriate. As was noted on page 96, it is not stock price (or index value) that matters to an investor, but rather stock return. The return is defined as the proportional change in price, or

$$r_i = \frac{p_i - p_{i-1}}{p_{i-1}} = \frac{p_i}{p_{i-1}} - 1.$$

Consider now prices in the natural log scale. Differencing this variable yields

$$\log p_i - \log p_{i-1} = \log \frac{p_i}{p_{i-1}}$$
$$= \log \left(1 + \frac{p_i}{p_{i-1}} - 1\right)$$
$$= \log(1 + r_i);$$

since the return r_i is usually close to 0, a Taylor series expansion yields $\log(1 + r_i) \approx r_i$ (Taylor series expansions will be discussed more fully in Section 11.3.1). That is, the differenced logged price series is roughly equal to the return series, and for this reason the differenced logged price is referred to as the **log return** in the finance and economics literature.

The slowly-decaying autocorrelations in Figure 5.11(b) are a common symptom of the need to difference a series. We can also see this in the following output of the regression of logged S&P large-cap price on lagged logged S&P large-cap price:

```
Coefficients:
                   Estimate Std.Error t value Pr(>|t|)
(Intercept)         0.02699  0.017421    1.55    0.122
lag.log.S.P.Large.cap 0.99619 0.002456  405.57   <2e-16 ***
---
Signif. codes:
   0 '***' 0.001 '**' 0.01 '*' 0.05 '.' 0.1 ' ' 1

Residual standard error: 0.01573 on 1256 degrees of freedom
   (1 observation deleted due to missingness)
Multiple R-squared: 0.9924,     Adjusted R-squared: 0.9924
F-statistic: 1.645e+05 on 1 and 1256 DF,  p-value: < 2.2e-16
```

The slope coefficient is very close to 1, highlighting that differencing the response variable (that is, using log returns) is an appropriate strategy. The residuals from this regression exhibit little autocorrelation [Figure 5.12(b)], but they now reflect two typical properties of stock returns: they are long-tailed relative to the normal distribution [Figure 5.12(c)], and they exhibit heteroscedasticity, with periods of low variability randomly alternating with periods of high variability [Figure 5.12(a)], with particularly high variability in late 2008. Both of these violations of least squares are consistent with certain time series models, such as ARCH, GARCH, and stochastic volatility models, which are often used to model stock returns. Such models are beyond the scope of this book, but see Gregoriou (2009) for further discussion of their use for modeling the volatility of stock returns.

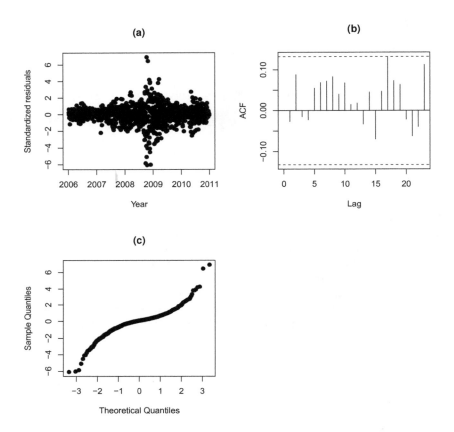

FIGURE 5.12 Residual plots for the logged large-cap index regression fit based on the lagged logged large-cap index. (a) Time series plot of standardized residuals. (b) ACF plot of residuals. (c) Normal plot of standardized residuals.

These results suggest regressing large-cap returns on small-cap and mid-cap returns, and output for that model is given below. Note that log returns (differenced logged prices) are used here, but results using actual returns are virtually identical.

```
Coefficients:
                       Estimate  Std.Error  t value  Pr(>|t|)
(Intercept)            -0.000142  0.000124    -1.14   0.25289
S.P.Small.cap.return   -0.095647  0.029382    -3.26   0.00116 **
S.P.Mid.cap.return      0.965094  0.031271    30.86   < 2e-16 ***
---
Signif. codes:
  0 '***' 0.001 '**' 0.01 '*' 0.05 '.' 0.1 ' ' 1
```

```
Residual standard error: 0.004411 on 1255 degrees of freedom
  (1 observation deleted due to missingness)
Multiple R-squared: 0.9215,     Adjusted R-squared: 0.9214
F-statistic:  7370 on 2 and 1255 DF,   p-value: < 2.2e-16
```

The regression is highly statistically significant, with 92% of the variability in large-cap returns accounted for by mid-cap and small-cap returns. The relationship is strongest between large- and mid-cap returns, but the negative slope of small-cap returns suggests that including them in a portfolio with mid-cap funds could provide useful diversification. The residuals exhibit autocorrelation, and the heteroscedasticity and long tails noted earlier (Figure 5.13), implying that more complex time series models are needed for these data. On the other hand, the runs test has $p = .31$, not pointing to autocorrelation.

Note that measures of strength of fit like R^2 and the overall F-statistic are not comparable between models using undifferenced and differenced response variables. The model on page 99 for logged large-cap index has $R^2 = 99.2\%$ while that for large-cap return above has $R^2 = 92.2\%$, yet the latter model clearly reflects a stronger relationship, as seen in the much smaller value of $\hat{\sigma}$ (these values *are* comparable, since the model for log return is a special case of a model for logged price that uses lagged logged price as a predictor and sets the slope to 1).

5.4.5 GENERALIZED LEAST SQUARES (GLS): THE COCHRANE-ORCUTT PROCEDURE

All of the methods discussed in the previous section fundamentally change the regression model being fit, whether it is by adding predictors (representing time trends, seasonal effects, or the lagged response) or changing the response completely (through differencing). Sometimes the original relationship hypothesized is the specific one of interest, and converting to a question about differences (for example) is not desired. The problem is then that OLS is an inappropriate criterion to use to fit the model, since the presence of autocorrelation is a violation of assumptions.

A solution to this problem is to use the "correct" criterion; that is, the one for which the autocorrelation present is assumed. This defines the **generalized least squares (GLS)** criterion, of which OLS is a special case. Equivalently, the idea is to transform the target and predictor variables so that the new variables satisfy a linear relationship based on the same parameters but satisfying the usual regression assumptions, and then use OLS to estimate those parameters. GLS estimation is not in general available in statistical software, but it turns out that for one particular type of autocorrelation it can be easily fit using OLS software. This is the essence of the **Cochrane-Orcutt procedure**, introduced by Cochrane and Orcutt (1949). We describe the algorithm for the single-predictor case, but it generalizes in a straightforward way to multiple predictors.

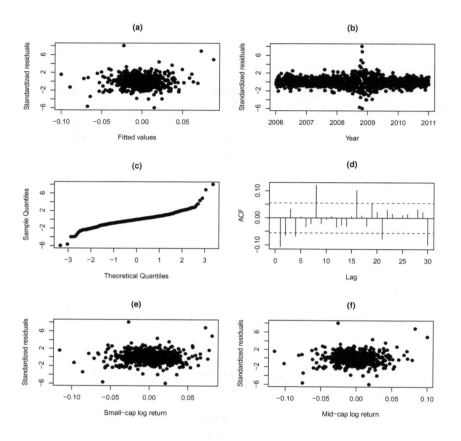

FIGURE 5.13 Residual plots for the large-cap log return regression fit based on small-cap log return and mid-cap log return. (a) Plot of standardized residuals versus fitted values. (b) Time series plot of standardized residuals. (c) Normal plot of standardized residuals. (d) ACF plot of residuals. (e) Scatter plot of standardized residuals versus small-cap log returns. (f) Scatter plot of standardized residuals versus mid-cap log returns.

The Cochrane-Orcutt procedure provides a GLS fit assuming the errors follow an AR(1) process, as in (5.1),

$$y_i = \beta_0 + \beta_1 x_i + \varepsilon_i, \quad \varepsilon_i = \rho \varepsilon_{i-1} + z_i.$$

Now, consider the transformation to

$$y_i^* = y_i - \rho y_{i-1}.$$

Substituting into the regression model yields

$$
\begin{aligned}
y_i^* &= y_i - \rho y_{i-1} \\
&= \beta_0 + \beta_1 x_i + \varepsilon_i - \rho(\beta_0 + \beta_1 x_{i-1} + \varepsilon_{i-1}) \\
&= \beta_0(1 - \rho) + \beta_1(x_i - \rho x_{i-1}) + \varepsilon_i - \rho \varepsilon_{i-1} \\
&= \beta_0(1 - \rho) + \beta_1(x_i - \rho x_{i-1}) + z_i \\
&\equiv \beta_0^* + \beta_1 x_i^* + z_i,
\end{aligned}
$$

where $\beta_0^* = \beta_0(1 - \rho)$ and $x_i^* = x_i - \rho x_{i-1}$, and the substitution in the fourth displayed line is based on (5.1). Thus, the regression of y_i^* on x_i^* provides estimates of β_0^* and β_1 that are appropriate (that is, they are the GLS estimates), since the errors z_i for the constructed regression are independent and identically normally distributed. We are not interested in β_0^*, but can get an estimate of β_0 by dividing the estimate of β_0^* by $1 - \rho$.

The Cochrane-Orcutt procedure is thus as follows:

1. Determine an estimate of ρ. A good one is the entry for lag 1 in the ACF plot of the residuals from an OLS fit (call it $\hat\rho$).

2. Form the transformed variables $y_i^* = y_i - \hat\rho y_{i-1}$ and $x_i^* = x_i - \hat\rho x_{i-1}$ (do this for each of the predicting variables in a multiple regression).

3. Fit the regression of y_i^* on the x_i^*'s using OLS. The slope estimates are left alone; the constant term estimate is adjusted using $\hat\beta_0 = \hat\beta_0^*/(1 - \hat\rho)$. A rough 95% prediction interval is $\hat y \pm 2\tilde\sigma$, where $\tilde\sigma = \hat\sigma/\sqrt{1 - \hat\rho^2}$ and $\hat\sigma$ is the standard error of the estimate from the Cochrane-Orcutt fit.

It is important to remember that the Cochrane-Orcutt procedure is merely a computational trick that allows a generalized least squares analysis using ordinary least squares programs. There is *no* physical meaning to y^* or x^*; they are merely tools that are used to get the GLS fit. However, since the Cochrane-Orcutt regression mimics that GLS fit, the usual measures of fit (R^2, F, t), residual plots, and regression diagnostics from the Cochrane-Orcutt fit can be interpreted in the usual way, since they are the appropriate ones from a GLS fit. The Cochrane-Orcutt procedure is not appropriate if a lagged version of the response variable is being used as a predictor.

A variation on the Cochrane-Orcutt procedure is the Prais-Winsten procedure (Prais and Winsten, 1954), which replaces x_1^* and y_1^* (which are missing when using Cochrane-Orcutt) with $x_1\sqrt{1 - \hat\rho^2}$ and $y_1\sqrt{1 - \hat\rho^2}$, respectively. Typically the results of the two approaches are very similar. In addition, each procedure can be iterated, by successively substituting the new estimates of β into the appropriate formulas, although usually if that is necessary a different approach should probably be tried.

5.4.6 EXAMPLE – TIME INTERVALS BETWEEN OLD FAITHFUL ERUPTIONS

A geyser is a hot spring that occasionally becomes unstable and erupts hot water and steam into the air. The Old Faithful Geyser at Yellowstone National Park in Wyoming is probably the most famous geyser in the world. Visitors to the park try to arrive at the geyser site to see it erupt without waiting too long; the name of the geyser comes from the fact that eruptions follow a relatively stable pattern. The National Park Service posts predictions of when the next eruption will occur at the Old Faithful Visitor Center and online. Thus, it is of interest to understand and predict the time interval until the next eruption.

The mechanism by which a geyser works suggests how the time to the next eruption might be predicted. Geysers occur near active volcanic areas, with about half of the roughly 1000 geysers in the world being in Yellowstone National Park. The eruption of a geyser comes from surface water working its way downwards through volcanic rock until it hits magma (molten rock). The combination of pressure and high temperatures heats the water far above the usual boiling temperature. The eventual boiling of the water results in superheated water and steam spraying out through the plumbing system of constricted fractures and fissures in the rock.

The observed duration of an eruption is subject to various random effects, and at Old Faithful can vary between roughly 90 seconds and five minutes. If an eruption turns out to be relatively short, less of the superheated water is sprayed out, meaning that it will tend to take less time for the remaining water (when mixed with new cold surface water) to be heated to boiling point. On the other hand, in a long eruption most of the heated water is lost, meaning that it will take longer until the next eruption. Thus, the duration of the previous eruption should be directly related to the time interval until the next eruption, with shorter time intervals following shorter eruptions and longer time intervals following longer eruptions. This suggests using regression to try to predict the time interval until the next eruption (and hence the time at which it will occur) from the duration of the previous eruption.

Figure 5.14, a plot of time interval to the next eruption versus duration of the previous eruption (both in minutes), shows that this is a reasonable idea. It is based on a sample of 222 eruption duration and following inter-eruption times taken during August 1978 and August 1979, as given in Weisberg (1980). It should be noted that the eruption behavior of the geyser has changed since this time, in particular because of the magnitude 6.9 earthquake that occurred less than 200 miles away at Borah Peak, Idaho, on October 29, 1983, so this analysis does not necessarily apply today. It can be seen that as expected there is a positive relationship between duration time and time to the next eruption. The following regression output summarizes the relationship.

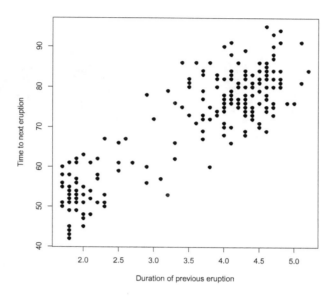

FIGURE 5.14 Scatter plot of the time interval to next eruption versus the duration of the previous eruption for eruptions of The Old Faithful Geyser.

```
Coefficients:
            Estimate Std. Error t value Pr(>|t|)
(Intercept) 33.9668     1.4279   23.79   <2e-16 ***
Duration    10.3582     0.3822   27.10   <2e-16 ***
---
Signif. codes:
  0 '***' 0.001 '**' 0.01 '*' 0.05 '.' 0.1 ' ' 1

Residual standard error: 6.159 on 220 degrees of freedom
Multiple R-squared: 0.7695,     Adjusted R-squared: 0.7685
F-statistic: 734.6 on 1 and 220 DF,   p-value: < 2.2e-16
```

There is clearly a strong relationship between the duration of the previous eruption and the time until the next eruption, with each additional minute's duration of the previous eruption associated with an estimated expected increase of 10.4 minutes in the time to the next eruption. Residual plots (Figure 5.15) suggest heteroscedasticity (with higher variability for longer fitted time intervals between eruptions and in the second half of the data), and also significant negative autocorrelation at the first lag ($\hat{\rho} = -.255$, $d = 2.55$, $p < .0001$). Negative autocorrelation is less common than positive autocorrelation, particularly for economic data, where cyclical behavior is consistent with positive autocorrelation. In this situation, however, the geyser erup-

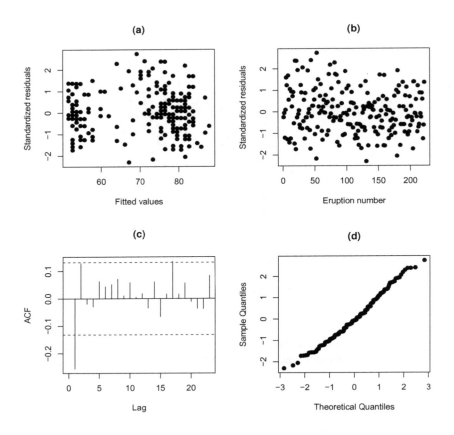

FIGURE 5.15 Residual plots for the Old Faithful time interval until the next eruption OLS regression fit based on duration of previous eruption. (a) Plot of standardized residuals versus fitted values. (b) Time series plot of standardized residuals. (c) ACF plot of residuals. (d) Normal plot of standardized residuals.

tion process makes negative autocorrelation reasonable. If an eruption takes longer to occur than expected (a positive error from the point of view of the regression model), the water will be heated to a higher than usual temperature; however long the eruption then is, the hotter water that remains will result in it taking less time than usual to heat all of the water in the geyser to the boiling point, meaning that it will take less time to the next eruption than expected (that is, a negative error is more likely to follow the positive error).

With the relatively small estimated autocorrelations it is difficult to assess the appropriateness of an AR(1) assumption for the errors, but as the first three estimated autocorrelations decrease and change signs in the appropriate way (from negative to positive to negative) a Cochrane-Orcutt GLS fit is not unreasonable. The corresponding regression output is as follows:

```
Coefficients:
            Estimate Std. Error t value Pr(>|t|)
(Intercept)   45.542       1.918   23.75   <2e-16 ***
Durationstar   9.711       0.418   23.23   <2e-16 ***
---
Signif. codes:
   0 '***' 0.001 '**' 0.01 '*' 0.05 '.' 0.1 ' ' 1

Residual standard error: 5.933 on 219 degrees of freedom
   (1 observation deleted due to missingness)
Multiple R-squared: 0.7113,     Adjusted R-squared:  0.71
F-statistic: 539.6 on 1 and 219 DF,  p-value: < 2.2e-16
```

The regression relationship is still strong, with estimated regression

$$\widehat{\text{Interval}} = 36.29 + 9.711 \times \text{Duration}$$

(after adjusting the intercept), so the GLS estimated intercept is slightly higher than the OLS one while the estimated slope is slightly lower. Residual plots (Figure 5.16) still indicate heteroscedasticity, but the autocorrelation appears to have been addressed ($d = 2.05$, and the runs test has $p = .21$). Regression diagnostics do not indicate any problems (the largest value of h_{ii} is .021 < $(2.5)(1 + 1)/221 = .023$, and the largest Cook's D is .02). The adjusted estimate of the standard deviation of the errors is $\tilde{\sigma} = 5.933/\sqrt{1 - (-.256)^2} = 6.138$, so a rough 95% prediction interval would be being able to predict the time to the next eruption to within $\pm(2)(6.14) = \pm12.3$ minutes. A rough 90% interval would correspond to $\pm(1.65)(6.14) = \pm10.1$ minutes, and the National Park Service reports on its Old Faithful Geyser prediction web site (http://www.nps.gov/yell) that their prediction model "has proven to be accurate, plus or minus 10 minutes, 90% of the time."

Note that since these data have gaps corresponding to separate days, several cases should actually be considered missing in the Cochrane-Orcutt fit, since the lagged duration and interval are not known for those cases. If this is done the results do not change appreciably.

5.5 Summary

In this chapter we have examined the effects of autocorrelation and various ways to identify it. We have also looked at simple ways of dealing with it within the context of least squares estimation, including accounting for time trends and seasonal effects, lagging and differencing series to account for the effects of earlier time periods and to address nonstationarity, and applying the Cochrane-Orcutt procedure to determine a GLS fit that is optimal for AR(1) errors.

As we have seen from the analyses in this chapter (where there is evidence for further autocorrelation effects), we have only scratched the surface

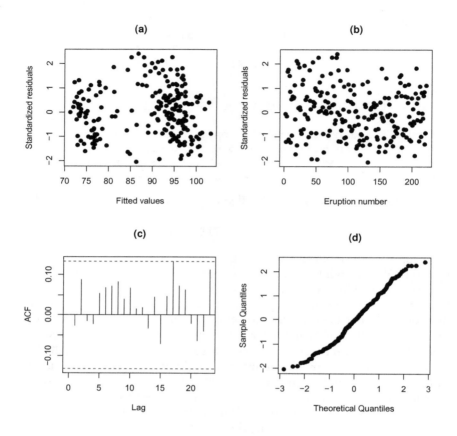

FIGURE 5.16 Residual plots for the Old Faithful time interval until the next eruption Cochrane-Orcutt GLS regression fit based on the duration of previous eruption. (a) Plot of standardized residuals versus fitted values. (b) Time series plot of standardized residuals. (c) ACF plot of residuals. (d) Normal plot of standardized residuals.

of time series modeling. Deseasonalizing can require more complicated approaches than simple constant shift models, and time series models more complicated than AR(1) (such as autoregressive/moving average [ARMA], ARCH, GARCH, and stochastic volatility models) can be appropriate. Despite this, the simple approaches discussed here will often account for a good deal of the autocorrelation present in regression errors, and can also serve as preliminary steps of an analysis that remove large-scale effects to allow subtler aspects of the data to emerge.

KEY TERMS

AR(1) process: A time series process, typically assumed for the errors when used in regression modeling, that is defined by $\varepsilon_i = \rho\varepsilon_{i-1} + z_i$, where z_i is a Gaussian (normal) random variable with mean 0, and $|\rho| < 1$.

Autocorrelation: The correlation that exists between the observations in time series data. The correlation between x_i and x_{i-r} is called the rth order autocorrelation.

Autocorrelation function (ACF): A plot of the estimated autocorrelation against the order (lag) of the autocorrelation.

Cochrane-Orcutt procedure: An algorithmic procedure used to construct the generalized least squares fit for a linear model to regression data assumed to have errors that follow an AR(1) process using ordinary least squares software. This is accomplished by working with transformed data.

Differenced variable: A variable generated by taking differences of adjacent observations; that is, for example, $y_2 - y_1, y_3 - y_2, \ldots, y_n - y_{n-1}$.

Durbin-Watson statistic: A statistic for testing the presence of AR(1) autocorrelation in the errors from a regression model fitting. Values close to 2 show absence of significant autocorrelation.

Lagged variable: A constructed variable whose ith value is an earlier value of an original series, usually the $(i-1)$st value. Using the lagged version of a response variable as a predictor in a regression can often help account for autocorrelation in the errors.

Log return: The first difference of the natural log of a stock price series. It is approximately equal to the proportional change in the price (the return).

Nonstationary time series: A time series for which the joint distribution of $\{x_i, x_{i-1}, \ldots, x_{i-m}\}$ is dependent on i for some value of m. A stationary process is one for which the joint distribution does not depend on i, which implies that the mean and variance of the series does not change over time.

Runs test: A nonparametric test used to assess whether a sequence of observations is random. The test is based on the number of runs of consecutive observations above or below a specified value, and significance is determined based on a permutation distribution for small samples and a normal approximation for large samples. In the regression context the test is typically based on runs of positive and negative residuals.

Seasonal effect: A recurring pattern in a time series linked to effects driven by the underlying annual cycle of the series. This can be reflected in autocorrelations at lags that are multiples of 4 in quarterly data, multiples of 12 in monthly data, multiples of 52 in weekly data, and so on.

Time trend: An general increasing or decreasing pattern in a time series, often reflecting smooth changes from population growth or inflation.

PART THREE

Categorical Predictors

Analysis of Variance

6.1   Introduction

In the regression models examined so far, both the target and predicting variables have been continuous, or at least effectively continuous — with one exception. The analysis of the pooled / constant shift / full model hierarchy in Section 2.4 recognized that the existence of two well-defined subgroups in the data could have predictive power for the target variable. That is, a *categorical* predicting variable taking on the values 0 and 1 could be used to address the effect of being in one or the other subgroup.

A natural question is to wonder if this can be generalized to more than two groups. For example, does knowing the educational level of a person (not a high school graduate, high school graduate, college graduate, or post-graduate degree) have predictive power for their annual salary? Is the return on a stock related to the industry group of the company? Do video games of different types have different expected sales? This is a regression question, but a special kind of regression question; in this context, saying that group membership has predictive power for the target is the same as saying that the average value of the target is different for different groups. That is, this is a question of the *comparison of means*.

In this and the following chapter we examine how regression models can be extended to allow for categorical predictors that identify multiple groups (or multiple levels of a variable). In this chapter we focus on models with only categorical predictors, which are termed **analysis of variance (ANOVA) models**. We start with one categorical predictor, **one-way ANOVA**, and then generalize this to **two-way ANOVA**. Discussion of the methodology used to fit such models, including the two different ways of coding them as regression models, follows. The important **multiple comparisons** problem is investigated, in which the goal is to compare the different groups to each other to determine which are actually different in terms of expected response. Since nonconstant variance often occurs for ANOVA data, with different groups having different variability of the errors, we then discuss **weighted least squares**, the generalization of ordinary least squares designed for this situation.

6.2   Concepts and Background Material

6.2.1 ONE-WAY ANOVA

Consider the simplest situation of one categorical predicting variable that takes on K values. The one-way ANOVA model is

$$y_{ij} = \mu + \alpha_i + \varepsilon_{ij}, \; i = 1, \ldots, K, \; j = 1, \ldots, n_i, \tag{6.1}$$

where y_{ij} is the value of y for the jth member of the ith group, μ is an overall level (roughly corresponding to the overall mean), α_i is the effect of being in the ith group, ε_{ij} is the error term, and n_i is the number of observations that fall in the ith group.

The α terms represent the difference in $E(y)$ that comes from being in any particular group relative to an overall level, since $E(y) = \mu + \alpha_i \equiv \mu_i$ for all observations in group i. It is natural to say that $\alpha_i = 0$ for all i if there is no difference between groups, but this requires a little more in terms of technical detail. Say $E(y) = 50$ for all groups, corresponding to no effect related to the categorical predictor. While this is obviously satisfied by $\mu = 50$ and $\alpha_1 = \cdots = \alpha_K = 0$, it is also satisfied by $\mu = 40$ and $\alpha_1 = \cdots = \alpha_K = 10$; indeed, there are an infinite number of possibilities that are consistent with this condition. For this reason, an additional constraint must be imposed on (6.1), that

$$\sum_{i=1}^{K} \alpha_i = 0.$$

With this additional constraint, it is guaranteed that a situation with no group effect will be modeled with $\alpha = 0$.

It is clear that the one-way ANOVA model is linear in its parameters, and is thus just a linear regression model based on specially-constructed predictors. This is discussed in Section 6.3.1.

6.2.2 TWO-WAY ANOVA

Consider now a situation with two categorical predictors, one (arbitrarily termed rows) with I levels and the other (arbitrarily termed columns) with J levels. This usage is based on the possibility of representing the $I \times J$ group means in the form of a table, and we will also sometimes refer to the combination of row level i and column level j as the (i, j)th cell. A simple generalization of model (6.1) is to add a set of parameters that correspond to the group effects of the second predictor. The model is then

$$y_{ijk} = \mu + \alpha_i + \beta_j + \varepsilon_{ijk}, \ i = 1, \ldots, I, \ j = 1, \ldots, J, \ k = 1, \ldots, n_{ij}, \quad (6.2)$$

where y_{ijk} is the value of y for the kth member of the (i, j)th group, μ is an overall level, α_i is the row effect for the ith level of the row variable, β_j is the column effect for the jth level of the column variable, ε_{ijk} is the error term, and n_{ij} is the number of observations that fall in the (i, j)th group. Just as was true for model (6.1), in order to make this model identifiable the additional constraints

$$\sum_{i=1}^{I} \alpha_i = \sum_{j=1}^{J} \beta_j = 0$$

are required. The main effects $\boldsymbol{\alpha}$ and $\boldsymbol{\beta}$ have a similar interpretation to the effect $\boldsymbol{\alpha}$ in the one-way ANOVA model (6.1), except in an analogous way to slope coefficients in a multiple regression versus a simple regression. In (6.2) each parameter corresponds to an effect holding the level of the other variable fixed. Thus, the model implies that (for example) the difference in expected response between an observation with row level i and one with row level i' equals $\alpha_i - \alpha_{i'}$ no matter which column level the observations come from, as long as they come from the same column level. Thus, the notion of a (single) overall row effect and (single) overall column effect is meaningful in this context.

It is easy to imagine, however, a situation where the effect on y of being in one row category versus another row category differs depending on the column category (or equivalently, the effect of being in one column category versus another column category differs depending on the row category). This is an interaction effect in the same way that the existence of different slopes for a predictor depending on group membership discussed in Section 2.4 was an interaction effect. This corresponds to an extension of model (6.2) to

$$y_{ijk} = \mu + \alpha_i + \beta_j + (\alpha\beta)_{ij} + \varepsilon_{ijk}, \ i = 1, \ldots, I, \ j = 1, \ldots, J, \ k = 1, \ldots, n_{ij}, \tag{6.3}$$

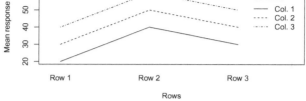

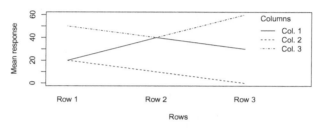

FIGURE 6.1 Interaction plots based on expected responses in situations without (top plot) and with (bottom plot) an interaction effect.

where the $(\alpha\beta)$ terms correspond to the existence of different row (column) effects for different columns (rows). The constraints

$$\sum_{i=1}^{I}(\alpha\beta)_{ij} = \sum_{j=1}^{J}(\alpha\beta)_{ij} = 0$$

make the model identifiable, and clearly

$$(\alpha\beta)_{11} = \cdots = (\alpha\beta)_{IJ} = 0$$

corresponds to a lack of the presence of an interaction effect.

The $(\alpha\beta)$ parameters themselves do not necessarily lend themselves to easy description of how the simple additivity of (6.2) is violated, but such a description is easily available through the use of an **interaction plot**. In such a plot the observed group means for each (i, j) combination are plotted in the form of separate lines connecting the means for each row, or alternatively separate lines connecting the means for each column. If model (6.2) holds (that is, there is no interaction effect), the differences between expected response values for each row (column) are the same for each column (row), implying that the lines in the plot corresponding to each row (column) will be parallel.

This is illustrated in the top plot of Figure 6.1. The relative position of the lines for the different columns shows that the column effect (which is the same for all rows) corresponds to increasing expected response going from levels 1 to 2 to 3, and the shape of those lines shows that the row effect (which is the same for all columns) corresponds to lowest expected response for level 1 and highest expect response for level 2.

This can be contrasted with the bottom plot in Figure 6.1, which is consistent with the existence of an interaction effect. In this case there is no single "row effect" or "column effect," since the row effect is different depending on the column level, and the column effect is different depending on the row level, resulting in lines that are not parallel in the plot. Specifically, in this case, the row effect for column level 1 implies an ordering in expected response from low to high of row levels $\{1, 3, 2\}$, for column level 2 an ordering of row levels $\{3, 2, 1\}$, and for column level 3 an ordering of row levels $\{2, 1, 3\}$.

Of course, in an interaction plot based on sample means one would expect to see some evidence of an apparent interaction even if the true model was (6.2) because of random fluctuation, so the interaction plot should not be viewed as a way to choose between models (6.2) and (6.3), but rather as a way to describe the interaction effect if it is decided that one exists. Choosing between the models would be done using the usual tools of regression model selection, which requires the ability to represent these models as regression models. This is discussed in the next section.

6.3   Methodology

6.3.1 CODINGS FOR CATEGORICAL PREDICTORS

Given that a categorical predictor with K levels is just a generalization from one with two levels, it seems natural that it can be incorporated into a regression model in a corresponding way using indicator variables. This is in fact the case. Just as was true when there were two levels of a categorical variable (such as Male/Female) and only one indicator variable was used to account for that effect, in the situation with K levels of a categorical variable, $K - 1$ indicator variables are used to account for that effect. Any one of the K possible indicator variables can be omitted, but the choice affects the interpretation of the slope coefficients.

Consider Table 6.1. The top table summarizes the implications of fitting a one-way ANOVA model (6.1) using indicator variables where the indicator for the Kth level is omitted; that is,

$$y_{ij} = \beta_0 + \beta_1 \mathcal{I}_{1ij} + \cdots + \beta_{K-1} \mathcal{I}_{K-1,ij} + \varepsilon_{ij}.$$

As is apparent from the table, in this situation the intercept β_0 equals the expected response when an observation comes from level K, the level for which the indicator was omitted. A slope coefficient β_m (say) then represents the difference in expected response between level m and level K for $m =$

TABLE 6.1 **Indicator variable codings for a one-way ANOVA fit omitting the variable for level** K **(top table) and level** 1 **(bottom table), respectively.**

Level	$\mathcal{I}_1$	$\mathcal{I}_2$		$\mathcal{I}_{K-2}$	$\mathcal{I}_{K-1}$	Expected response
Level 1	1	0		0	0	$\beta_0 + \beta_1$
Level 2	0	1		0	0	$\beta_0 + \beta_2$
$\vdots$			$\cdots$			$\vdots$
Level $K-2$	0	0		1	0	$\beta_0 + \beta_{K-2}$
Level $K-1$	0	0		0	1	$\beta_0 + \beta_{K-1}$
Level K	0	0		0	0	β_0

Level	$\mathcal{I}_2$	$\mathcal{I}_3$		$\mathcal{I}_{K-1}$	$\mathcal{I}_K$	Expected response
Level 1	0	0		0	0	β_0
Level 2	1	0		0	0	$\beta_0 + \beta_2$
Level 3	0	1		0	0	$\beta_0 + \beta_3$
$\vdots$			$\cdots$			$\vdots$
Level $K-1$	0	0		1	0	$\beta_0 + \beta_{K-1}$
Level K	0	0		0	1	$\beta_0 + \beta_K$

$1, \ldots, K-1$. That is, level K is a reference group, and the slope coefficients represent expected deviations from that reference group. If the omitted level was instead level 1, the model is

$$y_{ij} = \beta_0 + \beta_2 \mathcal{I}_{2ij} + \cdots + \beta_K \mathcal{I}_{Kij} + \varepsilon_{ij}.$$

As is summarized in the bottom table of Table 6.1, the reference group now becomes level 1, and all of the coefficients change accordingly (even though the expected responses do not change). The statistical significance of the effect of the categorical predictor (that is, the test of $\alpha = \mathbf{0}$) corresponds to a test of all of the slopes of the indicator variables equaling 0, and is thus tested using the usual overall F-test.

    This indicator variable formulation is a very reasonable one when there is a natural reference group in the context of the problem (for example, in a clinical trial situation where one level corresponds to the control group and the others are experimental treatment groups), but it is not an obvious choice when there is no natural reference group. Indeed, if there is no natural reference group examining slope coefficients based on indicator variables artificially forces one level into the role of "reference" group in a way that could be completely inappropriate. Further, an indicator variable-based fit does not correspond to the ANOVA model (6.1). To achieve that fit requires a different type of variable, called an **effect coding**. Effect codings look very much like indicator variables, except for the values corresponding to the level

TABLE 6.2 **Effect codings for a one-way ANOVA fit omitting the variable for level** K **(top table) and level** 1 **(bottom table), respectively.**

Level	E_1	E_2		E_{K-2}	E_{K-1}	Expected response
Level 1	1	0		0	0	$\beta_0 + \beta_1$
Level 2	0	1		0	0	$\beta_0 + \beta_2$
$\vdots$			$\cdots$			$\vdots$
Level $K-2$	0	0		1	0	$\beta_0 + \beta_{K-2}$
Level $K-1$	0	0		0	1	$\beta_0 + \beta_{K-1}$
Level K	-1	-1		-1	-1	$\beta_0 - \beta_1 - \cdots - \beta_{K-1}$ $(\equiv \beta_0 + \beta_K)$

Level	E_2	E_3		E_{K-1}	E_K	Expected response
Level 1	-1	-1		-1	-1	$\beta_0 - \beta_2 - \cdots - \beta_K$ $(\equiv \beta_0 + \beta_1)$
Level 2	1	0		0	0	$\beta_0 + \beta_2$
Level 3	0	1		0	0	$\beta_0 + \beta_3$
$\vdots$			$\cdots$			$\vdots$
Level $K-1$	0	0		1	0	$\beta_0 + \beta_{K-1}$
Level K	0	0		0	1	$\beta_0 + \beta_K$

for the variable that is omitted. While for indicator variables observations from the omitted level have the value 0 for every variable, for effect codings observations from the omitted level have the value -1 for every variable. It is simple to construct these variables from a set of indicator variables, as each effect coding variable just equals the indicator variable for the omitted level subtracted from the corresponding indicator variable.

This is laid out in Table 6.2. In the top plot the Kth level has its variable omitted, so any observations from that level take on the value -1 for every other variable (that is, the mth effect coding satisfies $E_m = \mathcal{I}_m - \mathcal{I}_K$), and the regression being fit is based on

$$y_{ij} = \beta_0 + \beta_1 E_{1ij} + \cdots + \beta_{K-1} E_{K-1,ij} + \varepsilon_{ij}.$$

This results in an expected response for level K of $\beta_0 - \beta_1 - \cdots - \beta_{K-1}$, rather than β_0 (as was the case for indicator variables). The average of the expected responses over all levels (the average of the values in the last column) is clearly β_0, which shows that in this formulation β_0 is an overall level of the response, not the expected response for a reference group. That is, it is μ from the ANOVA model (6.1). A slope coefficient β_m represents the difference between the expected response for level m and this overall level μ;

that is, $\beta_m = \alpha_m$ for $m = 1, \ldots, K - 1$ in (6.1). Finally, since the α terms sum to 0, the expected response for level K is implicitly $\mu + \alpha_K$, as implied by (6.1).

The key point here is that unlike when indicator variables are used, the interpretation of all of the coefficients using effect codings is the same no matter which level has its variable omitted. The bottom table in Table 6.2 demonstrates this, as all of the expected responses have the same form as in the top table, even though in this case it was the variable for the Kth level that was omitted. For this reason it is perfectly appropriate to report all $K + 1$ of the estimated coefficients (K slopes along with the intercept) to summarize the model, even though there are only K parameters being estimated; the constraint that makes the model identifiable ($\sum_i \alpha_i = 0$) is implicitly satisfied by the estimated coefficients. The existence of an effect is again tested using the overall F-test, which will be the same no matter which variable is omitted (and the same as that from using indicator variables).

The two-way ANOVA model without an interaction (6.2) is fit based on a regression using the $I - 1$ indicators or effect codings for rows and the $J - 1$ indicators or effect codings for columns. Statistical significance of the row (column) effect is based on the partial F-test comparing the fit using all of the variables to the one omitting the variables corresponding to rows (columns). In this context these tests are sometimes referred to as being based on "Type III" sums of squares, and are the same type of partial F-tests (and t-tests) discussed earlier. Fitted values are directly available from (6.2) and the estimated coefficients from the regression fit using effect codings, since

$$\widehat{E(y_{ij})} = \hat{\mu} + \hat{\alpha}_i + \hat{\beta}_j$$

(note that the β terms above refer to the column parameters in the two-way ANOVA model (6.2), not the parameters from the one-way ANOVA regression formulation summarized in Tables 6.1 and 6.2). The underlying components of these fitted values, $\hat{\mu} + \hat{\alpha}_i$ and $\hat{\mu} + \hat{\beta}_j$, are called the **least squares means** for rows and columns, respectively, as they estimate the expected response for each row (column) taking the column (row) effect into account.

The two-way ANOVA model with an interaction (6.3) also can be fit as a regression, but this requires the construction of indicator variables or effect codings for the interaction term. This is accomplished by calculating each of the $(I - 1)(J - 1)$ pairwise products of a row variable and a column variable (that is, $\mathcal{I}_{R1} \times \mathcal{I}_{C1}$, $\mathcal{I}_{R1} \times \mathcal{I}_{C2}$, etc., or $E_{R1} \times E_{C1}$, $E_{R1} \times E_{C2}$, etc., where the R (C) subscript refers to those being variables associated with row (column) levels. The statistical significance of the interaction effect is tested using the partial F-test comparing model (6.3) to model (6.2). Under model (6.3) the fitted value for any observation from the (i, j)th cell is the mean response for all observations from the (i, j)th cell, $\overline{Y}_{ij\cdot}$.

An interesting question concerns the order in which ANOVA models should be examined; that is, should an analyst "build up" from a model with only main effects to one including an interaction, or should one "simplify down" from a model with an interaction to one with only main effects. There are reasonable arguments in favor of either approach. A model that includes only main effects is analogous to a regression model on numerical predictors (where variable effects are constant given the values of the other predictors), and building up from that is consistent with the pooled model / constant shift model / full model hierarchy described in Section 2.4 (which will be generalized in the next chapter to allow for categorical variables with more than two levels). Adding an interaction effect then involves exploring whether it is possible to improve upon the simpler main effects model, and this can be tested using the corresponding partial F-test.

On the other hand, in many situations main effects are fairly obvious and therefore relatively uninteresting, and it is interaction effects that are the focus. In that situation discussing main effects is at best uninteresting and at worst potentially misleading, since in the presence of an interaction "the" row effect or "the" column effect is no longer meaningful (an interaction means that there are different row effects for different columns, and different column effects for different rows). So, for example, if the partial F-test for columns based a model using only main effects is insignificant, this could potentially encourage the analyst to think that the column variable has no relationship with the target variable and should be completely omitted from the analysis, when in fact it turns out that the row-column interaction effect is very strongly predictive. This possibility suggests starting with the model that includes the interaction effect, and using the partial F-test for it to decide if a simpler model with only main effects is adequate. Then, if that is the case, the main effects model could be fit, and row and column effect partial F-tests could be used to decide if the model can be simplified further.

What is certainly true is that when fitting a model that includes an interaction effect, the partial F-tests for the main effects do not correspond to testing meaningful hypotheses, and should not be examined. This does not mean that a general pattern of means for rows or columns must be ignored; for example, based on the bottom interaction plot of Figure 6.1 it would be reasonable for a data analyst to report that observations from column level 3 have generally the highest expected response, those from column level 1 have moderate expected response, and those from column level 2 have lowest expected response. What should not be done in a model that includes the interaction, however, is to appeal to the partial F-test for the column main effect to justify this statement. Further, a model that includes an interaction should include the corresponding main effects, since otherwise the simple interpretability of the interaction is lost (in fact, many statistical packages will refuse to fit a two-way ANOVA model that includes an interaction effect without including the corresponding main effects).

The situation where the number of observations n_{ij} is the same for all cells is termed a **balanced design**, and has several advantages over unbalanced designs. In a balanced design, the effects are orthogonal to each other; that is there is a perfect lack of collinearity between the codings that define one effect with those that define another effect. This implies that the variability accounted for by any effect (as measured by the change in residual sum of squares when including or excluding the effect) is the same, no matter which other effects are in the model. From a practical point of view, this means that the statistical significance of the row effect (for example) is (virtually) unchanged if the column effect is included or omitted from the model. If the design is very unbalanced, on the other hand, it can happen that a row effect (for example) is highly statistically significant when the model including an insignificant column effect is fit, but becomes insignificant if the apparently unneeded column effect is omitted from the model, which is counterintuitive and undesirable (but not surprising in the presence of collinearity).

Further, if the model is balanced, all observations have equal leverage. By contrast, in very unbalanced designs observations in cells with relatively few observations can have high leverage, thus potentially having a strong effect on the fitted ANOVA. In the most extreme situation, if a cell has only one observation ($n_{ij} = 1$) and the model includes an interaction effect, the leverage value for that observation will equal 1 (since one of the constructed effect codings will effectively be an indicator for that one observation). Omitting the observation is not a very desirable choice, since then it is impossible to fully estimate the interaction effect (it will not be possible to estimate the expected response for that combination of row and column levels). Indeed, some statistical packages will not fit ANOVA models with empty cells for this reason, and in that case it must be done by the analyst directly using regression on indicator variables or effect codings.

In the particular balanced design where $n_{ij} = 1$ for all cells the interaction effect cannot be fit. The reason for this is that in the model that includes the interaction effect there are IJ parameters in the model and IJ observations in the data, yielding identically zero residuals for all observations.

6.3.2 MULTIPLE COMPARISONS

Consider again the one-way ANOVA model (6.1). Given that it is believed that the categorical variable is meaningful as a predictor, it is natural to wonder which pairs of levels are different from each other. That is, if i and i' are two different levels of the variable, for which levels is $\alpha_i - \alpha_{i'}$ different from zero? This would seem to be testable in a fairly straightforward way using the partial F-test for the hypothesis $\alpha_i = \alpha_{i'}$, which is equivalent to the t-test for the slope of one level if the ANOVA is fit using indicator variables making the other level the reference level. The problem is that there is a multiple comparisons problem that comes from the many pairwise t-tests that are being calculated. Say there are K levels to the grouping variable. This implies that there are $C = \binom{K}{2} = K(K-1)/2$ different pairwise comparisons being made; for example, if $K = 10$ there are $C = 45$ comparisons being made. If each t-test was based on a .05 level of significance, and there were no differences between the groups, $(.05)(45) \approx 2$ of the tests would be statistically significant just by random chance.

Multiple comparisons procedures correct for this by controlling the *experimentwise* error rate. An experimentwise error rate of .05 says that in repeated sampling from a population where there is no difference between groups, only 5% of the time would *any* pair of groups be considered significantly different from each other. The broad area of multiple comparisons is beyond the scope of this book (for detailed discussion see, for example, Bretz et al., 2010), but some relatively simple methods are available for ANOVA models.

The most commonly-used approaches are the **Bonferroni correction** and **Tukey Honestly Significant Difference (HSD)** methods. The Bonferroni method argues that if the experimentwise error rate is desired to be α, each pairwise test should be done at an $\alpha/\mathcal{C}$ level, since that way the expected number of false positives is $(\alpha/\mathcal{C})(\mathcal{C}) = \alpha$. So, for example, for $K = 10$, each pairwise t-test would be conducted at a significance level of $.05/45 = .0011$. Equivalently, the Bonferroni-adjusted p-value for each t-test multiplies the usual p-value based on the t-distribution by $\mathcal{C}$ (and is 1 if that value is greater than 1). The Bonferroni correction is very general and very easy to apply, and usually does a good job of controlling the experimentwise error rate. Its only drawback is that it can sometimes be too conservative (that is, it does not reject the null as often as it should).

The Tukey method is a multiple comparisons method specifically derived for ANOVA multiple comparisons problems. This method determines significance and p-values of the pairwise t-statistics using the studentized range distribution, which is the exact correct distribution accounting for the multiple comparisons when the design is balanced. The Tukey test is less general than the Bonferroni correction, but is usually less conservative (particularly if the design is reasonably balanced).

Multiple comparisons methods for deciding which rows or columns are significantly different from each other generalize in a direct way to the two-way ANOVA model as long as the model does not include an interaction effect. Testing is not applicable in the presence of a fitted interaction, since the α and β terms are not interpretable in the presence of the $\alpha\beta$ terms in (6.3). It is possible to compare underlying cells to each other in a model with an interaction effect [that is, the mean response for cell (i, j) compared to that for cell (i', j')] using multiple comparisons methods, but this seems less useful from a practical point of view than examination of the interaction through an interaction plot.

An alternative approach to the multiple comparisons problem that has been investigated in the last 15 years is based on controlling the **false discovery rate**, which is the expected proportion of falsely rejected hypotheses among all rejected hypotheses. If all of the null hypotheses are true (that is, in the ANOVA context, all of the group expected responses are equal to each other) this is the same as the experimentwise rate controlled by the Bonferroni and Tukey methods, but when some of the null hypotheses are not true it is easier to reject the null, therefore making the test more sensitive and less conservative. This method is particularly attractive in the situation where a great many comparisons need to be made (such as in bioinformatics), since in that situation methods that control the experimentwise error rate can miss important differences that are actually present. Implementation of this method is not as straightforward as is implementation of the Bonferroni and Tukey methods, but it is becoming more generally available in software. See Strimmer (2008) for further discussion.

6.3.3 LEVENE'S TEST AND WEIGHTED LEAST SQUARES

The ANOVA situation (where there are by definition well-defined subgroups in the data) is one where heteroscedasticity (nonconstant variance) is common. Just as the responses for observations from different subgroups might have different means, the errors for observations from different subgroups might have different variances. This is a clear violation of the assumptions of ordinary least squares, with several negative impacts on OLS analyses.

1. The OLS estimates of the regression coefficients are inefficient. That is, on average, the OLS estimates are not as close to the true regression coefficients as is possible.

2. Inferential tests and intervals do not have the correct properties; confidence intervals do not have the correct coverage, and hypothesis tests do not have the correct significance levels.

3. Perhaps most importantly, predictions and prediction intervals are not correct. OLS assumes that the variance of all errors is the same (σ^2), which is reflected, for example, in the rough 95% prediction interval of $\hat{y}_i \pm 2\hat{\sigma}$. Clearly, if variances are different, so that $V(\varepsilon_i) = \sigma_i^2$, the appropriate interval is one of the form $\hat{y}_i \pm 2\hat{\sigma}_i$. So, for example, if the underlying variability of an observation (or a certain type of observation) is larger than that for another (type of) observation, the corresponding prediction intervals should be wider to reflect that. OLS-based intervals obviously do not have that property.

For these reasons it is important to try to identify potential nonconstant variance. As was noted in Section 1.3.5, residual plots are useful for this purpose, as varying heights of the point cloud can reflect different underlying variances (this also applies to side-by-side boxplots of residuals for categorical predictors, as can be seen in Figure 2.7 on page 46). More formal tests for nonconstant variance are also possible. A particularly simple one is **Levene's test**, in which the absolute value of the residuals from a regression or ANOVA fit is used as the response variable in a regression or ANOVA. Since observations with larger values of σ_i would be expected to have more extreme (and hence larger absolute) errors, evidence of a relationship between the absolute residuals and any predictors would be evidence of a relationship between the variance of the errors and those predictors. The Levene's test is the overall F-test based on the absolute residuals.

If nonconstant variance is actually a problem, there is a relatively straight-forward cure: **weighted least squares** (WLS). The idea behind WLS is the same as that behind the (Cochrane-Orcutt) GLS fit for a model exhibiting autocorrelated errors discussed in Section 5.4.5: transform the target and predicting variables to give a model with errors that satisfy the standard assumptions. Indeed, WLS is a special case of GLS. To keep the presentation simple, consider a simple regression model, although the discussion carries over directly to multiple regression and ANOVA situations. The regression model is

$$y_i = \beta_0 + \beta_1 x_i + \varepsilon_i,$$

but here with $V(\varepsilon_i) = \sigma_i^2$. We allow for nonconstant variance by setting $\sigma_i^2 = c_i^2 \sigma^2$. Dividing both sides of this equation by c_i gives

$$\frac{y_i}{c_i} = \beta_0 \left(\frac{1}{c_i} \right) + \beta_1 \left(\frac{x_i}{c_i} \right) + \frac{\varepsilon_i}{c_i}.$$

This can be rewritten

$$y_i^* = \beta_0 z_{1i} + \beta_1 z_{2i} + \delta_i,$$

where y_i^*, z_{1i}, z_{2i}, and δ_i are the obvious substitutions from the previous equation and $V(\delta_i) = \sigma^2$ for all i. Thus, OLS estimation (without an intercept term) of y^* on z_1 and z_2 gives fully efficient estimates of β_0 and β_1. From a conceptual point of view, the principle is that observations with errors with larger variances (larger c_i^2) have less information in them about the regression relationship, and therefore should be weighted less when estimating the regression parameters and conducting inference. This is achieved by estimating the regression parameters as the minimizer of the weighted residual sum of squares,

$$\sum_{i=1}^{n} w_i (y_i - \hat{y}_i)^2, \tag{6.4}$$

where $w_i = 1/c_i^2$ is the value of the weighting variable for the ith observation. Ordinary least squares is a special case of WLS with $w_i = 1$ for all i. In matrix notation, the resultant WLS estimates satisfy

$$\hat{\boldsymbol{\beta}} = (X'WX)^{-1}X'W\mathbf{y}, \tag{6.5}$$

where W is the diagonal matrix with ith diagonal element w_i. The weighted residual mean square based on (6.4) provides $\hat{\sigma}$ (which will be reported in any WLS output), and $\hat{\sigma}_i = \hat{\sigma}/\sqrt{w_i}$. The hat matrix has the form $H = X(X'WX)^{-1}X'W$, and all of the diagnostics from Chapter 3 carry over to WLS using this H (with ith diagonal element h_{ii}), and $\hat{\sigma}_i$ for the ith observation rather than $\hat{\sigma}$. In particular, standardized residuals and Cook's distances are defined this way based on (3.2) and (3.4), respectively.

The obvious difficulty is that σ_i^2 (or equivalently c_i^2) is unknown, and must be estimated. This is however easy to do in the ANOVA situation.

Consider a situation where there is a predictor defining K subgroups in the data. If the errors for all of the observations that come from group m (say) have the same variance, $\sigma^2_{[m]}$, then the weight for each of those observations would be (any constant multiple of) $1/\hat{\sigma}^2_{[m]}$. A choice that is then easily available is to use the inverse of the sample variance of the (standardized) residuals for the weight for members of group m. The situation where the variances are related to numerical predictors is more complicated, and will be discussed in Section 10.7. Technically, the fact that the weights are not fixed, but are rather estimated from the data, will have an effect on WLS-based inference, this effect is generally minor as long as the sample size is not very small.

Although generally speaking the appropriate choice of w_i is unknown and must be estimated from the data, sometimes that is not the case. Say that the response variable at the level of an individual follows the usual regression model,

$$y_i = \beta_0 + \beta_1 x_{i1} + \cdots + \beta_p x_{pi} + \varepsilon_i,$$

with $\varepsilon_i \sim N(0, \sigma^2)$ (this could of course be an ANOVA model). Imagine, however, that the ith observed response is actually an average $\overline{Y}_i$ for a sample of size n_i with the observed predictor values $\{x_{1i}, \ldots, x_{pi}\}$. The observed data thus actually satisfy

$$\overline{Y}_i = \beta_0 + \beta_1 x_{i1} + \cdots + \beta_p x_{pi} + \tilde{\varepsilon}_i,$$

where

$$V(\tilde{\varepsilon}_i) = V(\overline{Y}_i | \{x_{1i}, \ldots, x_{pi}\}) = \frac{\sigma^2}{n_i}.$$

An example of this kind of situation could be as follows. Say an analyst was interested in modeling the relationship between student test scores and (among other things) income. While it might be possible to obtain test scores at the level of individual students, it would be impossible to get incomes at that level because of privacy issues. On the other hand, average incomes at the level of census tract or school district might be available, and could be used to predict average test scores at that same level. This is not the same as predicting an individual's test scores from their particular income (since school districts are the units of study, not students), but could be useful from a policy point of view in terms of distributing resources over the various schools in the county, for example.

Clearly this is just a standard heteroscedasticity model, and WLS can be used to fit it. In this situation the weights do not need to be estimated at all; since $V(\tilde{\varepsilon}_i) = \sigma^2/n_i$, the weight for the ith observation is just n_i, with σ^2 estimated from the WLS residual mean square. That is, quite naturally, observations based on larger samples are weighted more heavily in estimating the regression coefficients. A similar situation is the presidential election data discussed in Section 2.4.1. For those data the response variable was the change in the percentage of votes cast for George W. Bush from 2000 to 2004.

These percentages are the empirical proportions of total voters who voted for Bush (multiplied by 100), and a binomial approximation implies that their variances are proportional to the total number of voters in the county. Figure 2.7 suggests nonconstant variance related to whether a county used electronic voting or not, but this is because larger counties were more likely to use e-voting, so the county voter turnout effect is confused with an e-voting effect.

6.3.4 MEMBERSHIP IN MULTIPLE GROUPS

Thus far we have focused on situations where each observation falls into one and only one category of a categorical variable. It is possible, however, that an individual could be a member of several of the groups defined by such a variable. For example, in a survey of college students, a respondent might want to identify themselves as being a member of several races or ethnicities, or as majoring in more than one field. Such responses would come from questions of the form "Check all that apply" when presented with a list of all race/ethnicity or majors groups.

The common approaches to this problem are to either add an extra category (such as "Multiracial" or "Double major," respectively), or to add extra categories for each of the possible combinations (such as "White and African American/Black," "White and Asian," and so on, or "Economics and Finances," "Economics and History," etc., respectively). Neither of these solutions is without problems. The former approach groups all combinations together, implying that all combinations of races or majors have the same expected relationship with the response variable, which seems unlikely. The latter approach is much more flexible, but is likely to result in many combinations with very few observations (there are $2^K - 1$ different membership possibilities), making inference difficult, and does not take advantage of any information from individuals who are members of only one group that could be relevant for individuals who are members of that group along with others.

It is possible to formulate a different approach using indicator variables or effect codings that has neither of these shortcomings, albeit at the cost of an additional assumption on the model. In order to handle multiple group membership, all that is required is to redefine the indicator variable for membership of individual i in group m to be $\mathcal{I}_{mi}^* = \mathcal{I}_{mi}/T$, where T is the total number of groups of which individual i is a member. Consider, for example, a categorical variable that takes on $K = 3$ levels. Table 6.3 summarizes the different possible combinations of group membership in this case, the adjusted indicator variable values for each combination, and the expected response if the variable for group 3 is omitted. The first three lines of the table show that the interpretation of the coefficients has not changed; β_0 is the expected response for the (omitted) reference group (i.e., μ_3), and β_1 and β_2 are the differences in expected response between group 1 or group 2 and group 3, respectively (i.e., $\mu_1 - \mu_3$ and $\mu_2 - \mu_3$, respectively).

This then implies an additional assumption about individuals who are members of multiple groups. Consider, for example, individuals who are

TABLE 6.3 Indicator variable codings for a one-way ANOVA fit with $K = 3$ omitting the variable for level 3 when multiple group membership is allowed.

Group membership			$\mathcal{I}_1^*$	$\mathcal{I}_2^*$	$\mathcal{I}_3^*$	Expected response omitting group 3
X			1	0	0	$\beta_0 + \beta_1$
	X		0	1	0	$\beta_0 + \beta_2$
		X	0	0	1	β_0
X	X		1/2	1/2	0	$\beta_0 + (\beta_1 + \beta_2)/2$
X		X	1/2	0	1/2	$\beta_0 + \beta_1/2$
	X	X	0	1/2	1/2	$\beta_0 + \beta_2/2$
X	X	X	1/3	1/3	1/3	$\beta_0 + (\beta_1 + \beta_2)/3$

TABLE 6.4 Effect codings for a one-way ANOVA fit with $K = 3$ omitting the variable for level 3 when multiple group membership is allowed.

Group membership			E_1^*	E_2^*	Expected response omitting group 3
X			1	0	$\beta_0 + \beta_1$
	X		0	1	$\beta_0 + \beta_2$
		X	-1	-1	$\beta_0 - \beta_1 - \beta_2$
X	X		1/2	1/2	$\beta_0 + (\beta_1 + \beta_2)/2$
X		X	0	$-1/2$	$\beta_0 - \beta_2/2$
	X	X	$-1/2$	0	$\beta_0 - \beta_1/2$
X	X	X	0	0	β_0

members of both groups 1 and 2. Their expected response satisfies

$$E(y) = \beta_0 + \frac{\beta_1 + \beta_2}{2} = \mu_3 + \frac{\mu_1 - \mu_3 + \mu_2 - \mu_3}{2} = \frac{\mu_1 + \mu_2}{2};$$

corresponding results hold for members of other combinations of groups. That is, this formulation assumes that the average response for an individual who is a member of multiple groups is the average of the expected responses for those groups.

The effect coding version of this formulation is summarized in Table 6.4. As before, each effect coding variable equals the indicator variable for the omitted level subtracted from the corresponding indicator variable, now based on the adjusted indicators $\mathcal{I}^*$ as given in Table 6.3. Once again the first three lines of the table show that the parameters have the same interpretation as before, consistent with the one-way ANOVA model (6.1). The last line directly shows that the expected response for individuals who are members of all three groups is the overall level μ, which of course equals the average of the responses for each of the three groups. Some simple algebraic manipulations shows that the other group memberships operate similarly, and equivalently to the use of (adjusted) indicator variables.

6.4 Example — DVD Sales of Movies

After-market revenues for movies from DVD sales have been a major profit center for studios since the introduction of the DVD in 1998, with revenues exceeding \$14 billion in 2004. In recent years the availability of films via digital download has cut into these revenues, making the ability to predict DVD sales even more important. Two characteristics of movies believed to be related to revenues (in both ticket sales and DVD sales) are the MPAA rating and the genre of the film.

This analysis is based on domestic DVD sales in millions of dollars for movies released on more than 500 screens in 2009. In order to avoid groups with small numbers of movies, four movies rated G, a documentary, and a musical are omitted from the analysis. Further, action and adventure movies are combined into the Action/Adventure genre, horror, romance, and thriller movies are included in the Drama genre, and romantic comedies are included in the Comedy genre. DVD sales are very long right-tailed, so logged (base 10) sales are used as the response variable. Five of the movies used in earlier analyses (Section 4.4) had missing values for DVD sales.

Figure 6.2 gives side-by-side boxplots separating logged DVD sales by the two categorical predictors. There is weak evidence for a rating effect, with PG-rated movies having highest sales, followed by PG-13-rated and then R-rated movies. There is stronger evidence of a genre effect, with action/adventure movies having highest revenues, followed by comedies and then dramas. There is also noticeable evidence of potential nonconstant variance, particularly related to genre.

We first fit a two-way ANOVA model that includes the interaction effect:

```
Response: Log.dvd
              Sum Sq  Df  F value  Pr(>F)
Rating        0.084    2   0.2520  0.77767
Genre         1.124    2   3.3653  0.03818 *
Rating:Genre  0.780    4   1.1676  0.32918
Residuals    18.197  109
---
Signif. codes:
  0 '***' 0.001 '**' 0.01 '*' 0.05 '.' 0.1 ' ' 1

Residual standard error: 0.4086 on 109 degrees of freedom
Multiple R-squared: 0.1536,    Adjusted R-squared: 0.09144
```

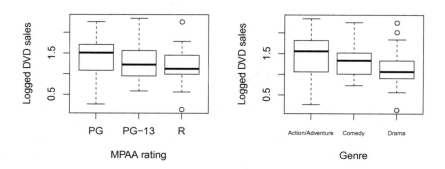

FIGURE 6.2 Side-by-side boxplots of logged DVD sales separated by MPAA rating and genre of movie.

It is apparent that the relationship is relatively weak, with R^2 only around 15%. The interaction effect is not close to statistical significance (recall that the F-tests for the main effects are not meaningful in the presence of the interaction). This suggests removing it and fitting the model with only the main effects, but boxplots of standardized residuals (Figure 6.3) show reasonably strong evidence of nonconstant variance related to genre (it is interesting to note that the evidence of nonconstant variance in logged DVD sales related to MPAA rating has disappeared in the residual plots).

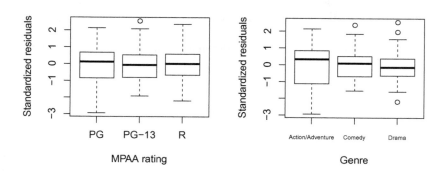

FIGURE 6.3 Side-by-side boxplots of standardized residuals from the two-way OLS ANOVA fit of logged DVD sales on MPAA rating, genre, and their interaction.

The Levene's test confirms that nonconstant variance is related to genre, but not MPAA rating.

```
Response: Abs.resid
            Sum Sq  Df  F value  Pr(>F)
Rating       0.388   2   0.5786  0.56235
Genre        2.430   2   3.6230  0.02985 *
Residuals   37.891 113
---
Signif. codes:
  0 '***' 0.001 '**' 0.01 '*' 0.05 '.' 0.1 ' ' 1
```

This suggests performing a WLS analysis, with the weights being based on genre. This is accomplished by determining the variance of the (standardized) residuals separated by genre group, and then weighting each observation from that group by the inverse of the variance. The resultant weights are $w = 1/1.5721$ for action/adventure movies, $w = 1/0.7245$ for comedies, and $w = 1/0.8692$ for dramas, respectively. This results in the following WLS-based ANOVA fit:

```
Response: Log.dvd
              Sum Sq  Df  F value  Pr(>F)
Rating         0.076   2   0.2360  0.79017
Genre          1.034   2   3.1949  0.04484 *
Rating:Genre   0.877   4   1.3552  0.25431
Residuals     17.639 109
---
Signif. codes:
  0 '***' 0.001 '**' 0.01 '*' 0.05 '.' 0.1 ' ' 1
```

```
Residual standard error: 0.4023 on 109 degrees of freedom
Multiple R-squared: 0.1543,      Adjusted R-squared: 0.09219
```

The entry for R^2 in the output above is worth further comment. Since the WLS estimates are based on minimizing the weighted residual sum of squares (6.4), the interpretation of R^2 as the observed proportion of variability accounted for by the regression is lost. In fact, the R^2 value typically reported in WLS output is a transformed version of the overall F-statistic for the ANOVA fit that uses the relationship between R^2 and the overall F-statistic for OLS fits, and thus has no physical interpretation. The WLS results are not very different from those of OLS, with the interaction effect still not close to statistical significance. Note, however, that now boxplots of standardized residuals (Figure 6.4) do not show evidence of nonconstant variance, and the Levene's test agrees (note that the Levene's test for a WLS fit must be based on standardized residuals, since unlike the ordinary residuals those take the weights into account).

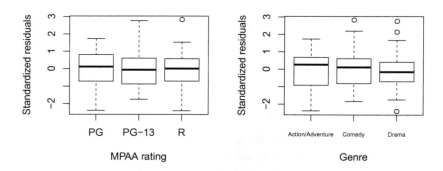

FIGURE 6.4 Side-by-side boxplots of standardized residuals from the two-way WLS ANOVA fit of logged DVD sales on MPAA rating, genre, and their interaction.

```
Response: Abs.resid
              Sum Sq  Df  F value  Pr(>F)
Rating        0.263    2   0.3739  0.6889
Genre         0.053    2   0.0747  0.9280
Residuals    39.711  113
---
Signif. codes:
  0 '***' 0.001 '**' 0.01 '*' 0.05 '.' 0.1 ' ' 1
```

It is now reasonable to omit the interaction effect from the model and examine the WLS fit based only on main effects. If the interaction effect had been needed, an interaction plot would be used to summarize its implications. The interaction plot is given in Figure 6.5 for completeness. The plot suggests that the genre effect of logged DVD sales for action/adventures and dramas are similar (with highest sales for PG-13 movies), while that for comedies is the opposite (with lowest sales for PG-13 movies), but the insignificant F-test for the interaction effect implies that this is likely to just be random fluctuation.

The WLS-based ANOVA fit that includes only the main effects suggests that only genre is predictive for logged DVD sales.

```
Response: Log.dvd
             Sum Sq  Df F value  Pr(>F)
Rating       0.3275   2  0.9994  0.37133
Genre        1.3232   2  4.0375  0.02025 *
Residuals   18.5162 113
---
```

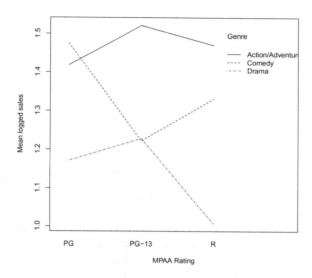

FIGURE 6.5 Interaction plot for logged DVD sales separated by MPAA rating and genre.

```
Signif. codes:
  0 '***' 0.001 '**' 0.01 '*' 0.05 '.' 0.1 ' ' 1

Residual standard error: 0.4048 on 113 degrees of freedom
Multiple R-squared: 0.1122,      Adjusted R-squared: 0.08078

--------------------


Response: Log.dvd
            Sum Sq  Df F value   Pr(>F)
Genre       2.0127   2  6.1415 0.002923 **
Residuals  18.8437 115
---
Signif. codes:
  0 '***' 0.001 '**' 0.01 '*' 0.05 '.' 0.1 ' ' 1

Residual standard error: 0.4048 on 115 degrees of freedom
Multiple R-squared: 0.0965,      Adjusted R-squared: 0.08079
```

Since the final model is a one-way ANOVA, the fitted value for any movie of a particular genre is just the mean response of movies of that genre. Thus, the fitted logged DVD sales for action/adventure movies is 1.471 (corresponding to a geometric mean sales of $10^{1.471} = \$29.6$ million), for comedies is 1.320 ($\$20.9$ million), and for dramas is 1.132 ($\$13.6$ million). Tukey multiple comparisons tests show that mean logged sales are statistically significantly different between action/adventure movies and dramas and between comedies and dramas, but not between action/adventure movies and comedies.

```
Multiple Comparisons of Means: Tukey Contrasts

Linear Hypotheses:
                         Estimate  Std.Err.  t val  Pr(>|t|)
Comedy = Action/Adventure  -0.1517  0.1024   -1.48  0.30083
Drama = Action/Adventure   -0.3390  0.1026   -3.30  0.00357 **
Drama = Comedy             -0.1873  0.0793   -2.36  0.05068 .
---
Signif. codes:
   0 '***' 0.001 '**' 0.01 '*' 0.05 '.' 0.1 ' ' 1
```

The output from the one-way ANOVA WLS fit gives $\hat{\sigma} = 0.4048$. Recalling that $\hat{\sigma}_i = \hat{\sigma}/\sqrt{w_i}$, this implies estimated standard deviations of the errors of $(.4048)\sqrt{1.5721} = .508$ for action/adventure movies, $(.4048)\sqrt{0.7245} = .345$ for comedies, and $(.4048)\sqrt{0.8692} = .377$ for dramas, respectively. This in turn implies exact 95% prediction intervals for logged DVD sales of $(0.452, 2.491)$ for action/adventure movies, $(0.628, 2.011)$ for comedies, and $(0.377, 1.888)$ for dramas, respectively. Unfortunately, these are far too wide to be useful in practice, since antilogging the ends of the intervals implies predictive ranges from roughly $2 to $4 million to upwards of $300 million (for action/adventure movies). This is not particularly surprising, given the observed weak relationship between logged DVD sales and genre. A more reasonable model would also include other predictors, including numerical ones (such as total gross revenues of the movie in theaters). Generalizing regression modeling to allow for the possibility of both numerical and categorical predictors is the topic of the next chapter.

Residual and diagnostic plots given in Figure 6.6 show that nonconstant variance is no longer a problem (note that the weighting is taken into account in the calculation of leverage values and Cook's distances). There are five marginal outliers ("The Blind Side," "The Hangover," and "The Twilight Saga: New Moon" did unusually well for their respective genres, while "Shorts" and "Sorority Row" did unusually poorly). An analysis after omitting these points (not given) does not change the results of the model selection process or the implications of the results, but does result in a stronger apparent relationship between logged DVD sales and genre, to the extent that mean logged DVD sales for comedies and dramas are now marginally statistically significantly different from each other.

6.5 Higher-Way ANOVA

ANOVA models generalize past one or two grouping variables, at the cost of some additional complications in the model. Consider the situation of

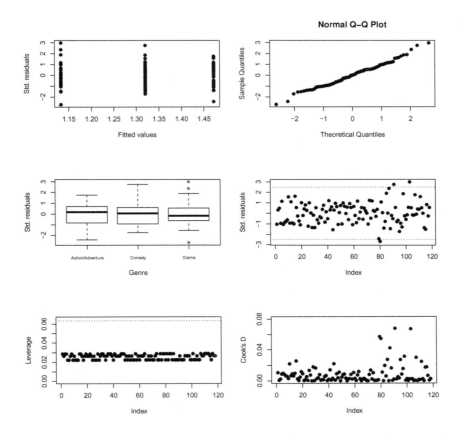

FIGURE 6.6 Residual and diagnostic plots for the WLS-based ANOVA fit of logged DVD sales on genre.

three grouping variables, sometimes generically referred to rows, columns, and layers. A full three-way ANOVA model would have the form

$$y_{ijk\ell} = \mu + \alpha_i + \beta_j + \gamma_k + (\alpha\beta)_{ij} + (\alpha\gamma)_{ik} + (\beta\gamma)_{jk} + (\alpha\beta\gamma)_{ijk} + \varepsilon_{ijk\ell}.$$

The main effects α, β, and γ are defined as they were before, as are the two-way interactions $(\alpha\beta)$, $(\alpha\gamma)$, and $(\beta\gamma)$. Just as a two-way interaction is defined as the presence of different main effects for one factor as the level of the second factor changes, a three-way interaction occurs if the two-way interaction of two factors differs depending on the level of the third factor. An interaction plot would identify this graphically as different patterns in interaction plots of rows and columns for different layers. This can be extended further to more potential grouping variables, although difficulties in interpretation make interactions that are higher order than three-way rarely used in practice.

The model is fit in the same way as the simpler ones were fit. The variables that define the three-way interaction are all of the products taking one effect coding each from those corresponding to rows, columns, and layers. As before, if only one observation per cell is available, the three-way interaction (being the highest order interaction) cannot be included in the model.

6.6   Summary

Categorical variables occur often in practice, and building models using them as predictors is often desirable. The use of indicator variables or effect codings makes fitting of these ANOVA models explicitly a linear regression problem, allowing all of the power of linear modeling to be used. The nature of categorical variables as ones that identify well-defined subgroups in the data means that nonconstant variance related to the existence of those subgroups is a common problem. For this reason, weighted least squares (WLS)-based fits of ANOVA data are often advisable. Multiple comparisons methods such as the Tukey HSD test and the use of Bonferroni corrections allow for the comparison of mean responses for multiple pairs of groups that take into account the potentially large number of comparisons being made.

There is no requirement that a regression relationship be based only on numerical predictors, or only on categorical ones. In the next chapter we discuss the natural generalization to the situation of a mixture of such variable types.

KEY TERMS

Balanced design: A data set in which the number of observations is the same for all combinations of group levels.

Bonferroni correction: A correction applied in a multiple testing situation designed to keep the overall level of significance of the set of tests, rather than the significance level of each test, at α. When C tests are performed, the significance level is taken to be α/C for each individual test, in order to ensure that the overall level of significance is no greater than α.

Effect coding: A variant of indicator variables that results in regression coefficients that represent deviations from an overall level when used to code factors in an analysis of variance model.

Interaction effect: In two-way classified data, a pattern wherein the row effect on an observation differs depending on the column in which it is located, and vice versa. In higher-order classified data, a pattern wherein a lower-order interaction effect differs depending on the level of a factor not in the interaction.

Interaction plot: A plot of each of the row effects separated by column level, or equivalently of each of the column effects separated by row level. Roughly parallel lines indicate the lack of an interaction effect.

Levene's test: A test for heteroscedasticity based on using the absolute value of the (standardized) residuals as the response variable in a regression or ANOVA fit.

Main effect: The effect on an observation related to the row or the column in which it occurs.

Multiple comparisons: The statistical inference problem in which (adjusted) group means are compared simultaneously in an analysis of variance model.

One-way analysis of variance (ANOVA): A methodology used to test the equality of means of several groups or levels classified by a single factor.

Two-way analysis of variance (ANOVA): A methodology used to test the equality of means of a several groups or levels classified by two factors. The analysis allows variation between means to be separated into row or column main effects, as well as an interaction effect.

Tukey's HSD test: A test that directly addresses the multiple comparisons problem by using the (correct) studentized range distribution for differences between sample means.

Weighted least squares (WLS): A generalization of ordinary least squares (OLS) that accounts for nonconstant variance by weighting observations with smaller (estimated) variances more heavily than those with larger (estimated) variances.

Analysis of Covariance

7.1   Introduction

The analysis of variance (ANOVA) models of the previous chapter are restrictive in that they allow only categorical predicting variables. **Analysis of covariance (ANCOVA)** models remove this restriction by allowing both categorical predictors (often called *grouping variables* or *factors* in this context) and continuous predictors (typically called *covariates*) in the model. This can be viewed as a generalization of the ANOVA models of Chapter 6 to models that include covariates, or a generalization of the models discussed in Section 2.4 based on indicator variables to models that allow for categorical variables with more than two categories.

7.2   Methodology

7.2.1 CONSTANT SHIFT MODELS

Conceptually ANCOVA models are quite straightforward, since they merely add additional numerical predictors to the constructed (effect coding) variables used in ANOVA modeling. If there is one grouping variable, for example, the model is

$$y_{ij} = \mu + \alpha_i + \beta_1 x_{1ij} + \cdots + \beta_p x_{pij} + \varepsilon_{ij}, \ i = 1, \ldots, K, \ j = 1, \ldots, n_i, \ (7.1)$$

where α_i is the corrected effect on y given membership in group i (corrected in the sense that the covariates $\mathbf{x}_1, \ldots, \mathbf{x}_p$ are taken into account, and subject to the usual constraint $\sum_i \alpha_i = 0$), and β_ℓ is the slope coefficient corresponding to the expected change in the response associated with a one unit change in x_ℓ given group membership and all of the other covariates are held fixed. This model is fit using $K-1$ effect codings to represent the grouping variable, along with the p covariates and the constant term.

There are several obvious hypotheses of interest based on this model:

1. Are there differences in expected level between groups (given the covariates)? This tests the null hypothesis

$$H_0 : \alpha_1 = \cdots = \alpha_K = 0.$$

The test used for this hypothesis is the partial F-test for the $K-1$ effect coding variables (that is, it is based on the residual sum of squares using all of the variables, and the residual sum of squares using only the covariates). Further analysis exploring which groups are significantly different from which (given the covariates) is meaningful here, and the multiple comparisons methods of Section 6.3.2 can be adapted to this model.

2. Do the covariates have any predictive power for y (given the grouping variable)? This tests the null hypothesis

$$H_0 : \beta_1 = \cdots = \beta_p = 0.$$

The test used for this hypothesis is the partial F-test for the covariates (that is, it is based on the residual sum of squares using all of the variables, and the residual sum of squares using only the effect codings).

3. Does a particular variable x_j provide any predictive power given the grouping variable and the other covariates? This tests the null hypothesis

$$H_0 : \beta_j = 0.$$

The test used for this hypothesis is the usual t-test for that covariate.

Note that since ANCOVA models are just regression models, all of the model selection approaches discussed in Section 2.3.1 apply here as well, although they are less commonly used in practice in this situation.

This model generalizes to more than one grouping variable as well. For two grouping variables, for example, the model is

$$y_{ijk} = \mu + \alpha_i + \beta_j + (\alpha\beta)_{ij} + \gamma_1 x_{1ijk} + \cdots + \gamma_p x_{pijk} + \varepsilon_{ijk}, \qquad (7.2)$$

which allows for two main effects (fit using effect codings for each grouping variable) and an interaction effect (fit using the pairwise products of the effect codings for the main effects), as well as the presence of covariates. The usual ANOVA hypotheses about the significance of main effects and the interaction effect are tested using the appropriate partial F-tests, as described in Sections 6.2.2 and 6.3.1.

7.2.2 VARYING SLOPE MODELS

Models (7.1) and (7.2) are constant shift models, in the sense that the only differences between the expected value of the target variable for a given set of covariate values between groups is one of shift, with the slopes of the covariates being the same no matter what group an observations falls in. This implies a natural question: might the slopes also be different for different levels of the grouping variable? Assume for simplicity that there is one covariate x in the model. A generalized model that allows for different slopes for different groups is

$$y_{ij} = \mu + \alpha_i + \beta_{1i}x_{ij} + \varepsilon_{ij}, \tag{7.3}$$

where β_{1i} is the slope of x for the ith group. If the interaction of the grouping variable and the covariate are entered as part of the general linear model (by including the pairwise products of the effect codings and the covariate), the partial F-test for this set of variables is a test of the hypothesis

$$H_0 : \beta_{11} = \cdots = \beta_{1K}$$

(this is often called a **test of common slope**). This is easily generalized to more than one covariate using the appropriate interaction terms. This can also be generalized to the situation with more than one grouping variable, but that is less common.

7.3   Example — International Grosses of Movies

Although domestic (U.S. and Canada) gross revenues of movies are the numbers routinely reported in the American news media, revenues from other countries can often outstrip domestic revenues and mean the difference between profit and loss. It is thus of interest to try to model international gross revenues. This analysis is based on revenues for movies released in the United States on more than 500 screens in 2009, with logged (base 10) domestic grosses and MPAA rating used to model logged (base 10) international grosses. Figure 7.1 shows that movies with higher domestic revenues tend to have higher international revenues, and movies rated G and PG tend to have higher international revenues than those rated PG-13 and R. There is also evidence of nonconstant variance, with movies with lower domestic revenues having higher variability.

We first fit a constant shift model:

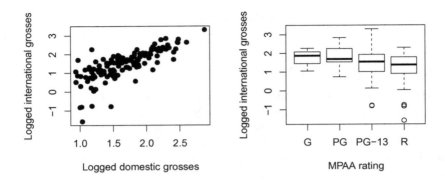

FIGURE 7.1 Plots for the 2009 international movie grosses data. (a) Plot of logged international gross versus domestic gross. (b) Side-by-side boxplots of logged international gross by MPAA rating.

```
Response: Log.international.gross
                       Sum Sq  Df  F value     Pr(>F)
(Intercept)             3.610   1  14.2648 0.0002468 ***
Log.domestic.gross     42.630   1 168.4463 < 2.2e-16 ***
Rating                  2.712   3   3.5722 0.0160807 *
Residuals              30.876 122
---
Signif. codes:
  0 '***' 0.001 '**' 0.01 '*' 0.05 '.' 0.1 ' ' 1


Coefficients:
                   Estimate Std. Error t value Pr(>|t|)
(Intercept)       -0.710076   0.188006  -3.777 0.000247 ***
Log.domestic.gross 1.385308   0.106737  12.979  < 2e-16 ***
Rating1            0.440935   0.192664   2.289 0.023822 *
Rating2            0.008748   0.098965   0.088 0.929705
Rating3           -0.182498   0.086866  -2.101 0.037707 *
---
Signif. codes:
  0 '***' 0.001 '**' 0.01 '*' 0.05 '.' 0.1 ' ' 1

Residual standard error: 0.5031 on 122 degrees of freedom
Multiple R-squared: 0.6087,     Adjusted R-squared: 0.5958
F-statistic: 47.44 on 4 and 122 DF,  p-value: < 2.2e-16
```

Logged domestic gross and MPAA rating account for roughly 60% of the variability in logged international gross. The standard error of the estimate of $\hat{\sigma} = 0.503$ implies that the model can predict international grosses to within a multiplicative factor of roughly 10, 95% of the time, clearly a large range. The coefficient for logged domestic gross (which is highly statistically significant) implies that, given MPAA rating, a 1% increase in domestic gross is associated with an estimated expected 1.39% increase in international gross. The three effect codings (`Rating1`, `Rating2`, and `Rating3`) refer to effects for movies rated G, PG, and PG-13, respectively, and their coefficients imply much higher grosses for G-rated movies than the other three types, given logged domestic gross. Since the coefficients must sum to 0, the implied estimated coefficient for R-rated movies is -0.267185. The rating effect, while much less strong than that of logged domestic gross, is also statistically significant.

Since there is no interaction term in this model, multiple comparison tests based on the Tukey method (Section 6.3.2) can be used to assess which rating classes are significantly different from each other given logged domestic gross.

```
Multiple Comparisons of Means: Tukey Contrasts

Linear Hypotheses:
            Estimate  Std. Error  t value  Pr(>|t|)
PG = G      -0.43219   0.27028    -1.599    0.3620
PG-13 = G   -0.62343   0.26165    -2.383    0.0774 .
R = G       -0.70812   0.26437    -2.679    0.0369 *
PG-13 = PG  -0.19125   0.11574    -1.652    0.3329
R = PG      -0.27593   0.12512    -2.205    0.1160
R = PG-13   -0.08469   0.10608    -0.798    0.8449
---
Signif. codes:
   0 '***' 0.001 '**' 0.01 '*' 0.05 '.' 0.1 ' ' 1
```

It can be seen that G-rated movies have significantly higher international grosses than do R-rated movies ($p = .037$, corresponding to $10^{.708} = 5.1$ times the grosses given logged domestic gross) and marginally significantly higher international grosses than do PG-13-rated movies ($p = .077$, corresponding to $10^{.623} = 4.2$ times the grosses given logged domestic gross). Note that these effects are larger than the corresponding marginal MPAA rating effects (for example, if logged domestic gross is omitted from the model, G-rated movies are estimated to have only 3.6 times the international gross than R-rated movies, not 5.1 times), showing that the poorer relative international performance of PG-13 or R-rated movies is ameliorated somewhat by their stronger relative domestic performance in these data.

A generalization of this model would be to model (7.3), allowing for different slopes for logged domestic gross for different MPAA rating classes. This involves adding the interaction between the categorical predictor and the covariate:

```
Response: Log.international.gross
                               Sum Sq  Df  F value    Pr(>F)
(Intercept)                    0.1170   1   0.4892   0.485630
Log.domestic.gross             5.0368   1  21.0565  1.113e-05  ***
Rating                         3.4105   3   4.7526   0.003633  **
Log.domestic.gross:Rating      2.4103   3   3.3588   0.021173  *
Residuals                     28.4652 119
---
Signif. codes:
  0 '***' 0.001 '**' 0.01 '*' 0.05 '.' 0.1 ' ' 1


Coefficients:
                          Estimate Std.Error  t val  Pr(>|t|)
(Intercept)               -0.25185    0.3601  -0.70    0.4856
Log.dom.gross              1.08408    0.2363   4.59  1.11e-05  ***
Rating1                    1.87935    0.9991   1.88    0.0624  .
Rating2                    0.24720    0.4511   0.55    0.5847
Rating3                   -0.67295    0.4019  -1.67    0.0967  .
Log.dom.gross:Rating1     -0.99180    0.6676  -1.49    0.1400
Log.dom.gross:Rating2     -0.09207    0.2793  -0.33    0.7423
Log.dom.gross:Rating3      0.32025    0.2572   1.25    0.2155
---
Signif. codes:
  0 '***' 0.001 '**' 0.01 '*' 0.05 '.' 0.1 ' ' 1

Residual standard error: 0.4891 on 119 degrees of freedom
Multiple R-squared: 0.6392,    Adjusted R-squared: 0.618
F-statistic: 30.12 on 7 and 119 DF,  p-value: < 2.2e-16
```

Adding the interaction effect does not increase the fit greatly, but the partial F-test of constant slope does imply that the slopes for logged domestic gross are statistically significantly different across MPAA ratings. Figure 7.2 indicates the different implications of the two models. The plot on the left gives the fitted lines relating logged international and logged domestic grosses when restricting the model to constant slope, while the plot on the right allows the slopes to be different. The interesting pattern emerges that as the potential audience for a movie shrinks (MPAA rating going from G to PG to PG-13 to R) the importance of domestic gross as a predictor of international gross grows. While for G-rated movies there is virtually no relationship between domestic and international grosses, a 1% increase in domestic grosses is associated with an estimated expected 1%, 1.4%, and 1.85% increase in international grosses for PG-, PG-13-, and R-rated movies, respectively. Given that the proportion of the audience that are adults grows going from G to R ratings, it appears that internationally adults are more sensitive to the domestic success of a movie than are children.

Residual plots given in Figure 7.3 highlight several violations of assumptions here, so these results cannot be viewed as definitive. There is appar-

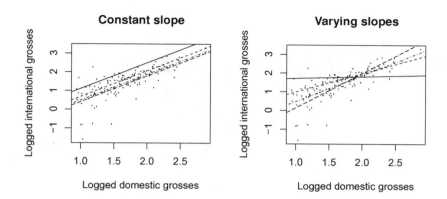

FIGURE 7.2 ANCOVA fits for the 2009 international movie grosses data. Solid line: G-rated movies. Short dashed line: PG-rated movies. Dotted-and-dashed line: PG-13-rated movies. Long dashed line: R-rated movies. (a) Fitted lines for the constant slope model. (b) Fitted lines for the varying slope model.

ent nonconstant variance, with movies with lower domestic revenues having higher variability in international revenues, and the residuals are long left-tailed. As was noted earlier, the situation where the variances are related to a numerical predictor will be discussed further in Section 10.7.

7.4 Summary

Analysis of covariance models represent the natural generalization of analysis of variance models to allow for numeric covariates. The simplest model is a constant shift (constant slope) model, but models that allow for varying slopes for different groups are also easily constructed. Since various models of interest are nested within each other, partial F-tests are a natural way to compare models, although information measures such as AIC_c also can be used for this purpose.

KEY TERMS

Analysis of covariance: A methodology used to analyze data characterized by several groups or levels classified by a single factor or multiple factors and numerical predictors. The analysis allows variation between group means to be separated into main effects and (potentially) interaction effects, as well as effects related to the numerical predictors (including the possibility of different slopes for different groups).

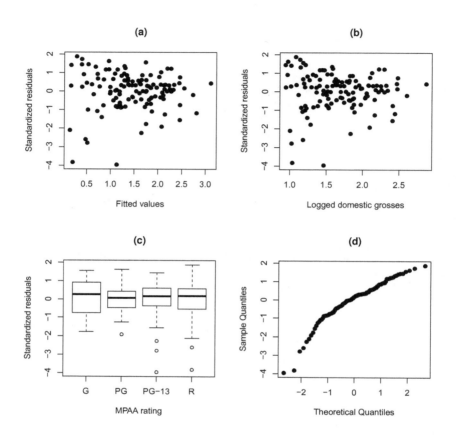

FIGURE 7.3 Residual plots for the varying slopes ANCOVA model for the 2009 international grosses data. (a) Plot of standardized residuals versus fitted values. (b) Plot of standardized residuals versus logged domestic gross. (c) Side-by-side boxplots of standardized residuals by MPAA rating. (d) Normal plot of standardized residuals.

Covariate: A numerical predictor in an analysis of covariance model.

PART FOUR

Other Regression Models

Logistic Regression

8.1   Introduction

All of the regression situations discussed thus far have been characterized by a response variable that is continuous, but the modeling of a categorical variable having two (or more) categories is sometimes desired. Consider a study of risk factors for cancer. From the health records of subjects data are collected on age, sex, weight, smoking status, dietary habits, and family's medical history. In this case the response variable is whether the person has lung cancer ($y = 1$), or does not have lung cancer ($y = 0$), and the question of interest is "What factors can be used to predict whether or not a person will have lung cancer?". Similarly, in financial analysis the solvency ("health") of a company is of great interest. In such an analysis the question of interest is "What financial characteristics can be used to predict whether or not a business will go bankrupt?".

In this chapter we will focus on data where the response of interest takes on the two values 0 and 1; that is, **binary data**. This will be generalized to more than two values in the next chapter. In this situation the expected response $E(y_i)$ is the conditional probability of the event of interest ($y_i = 1$, generically termed a success, with $y_i = 0$ termed a failure) given the values of the predictors. A linear model would be

$$E(y_i) \equiv \pi_i = \beta_0 + \beta_1 x_{1i} + \cdots + \beta_p x_{pi}.$$

It is clear that a linear least squares modeling of this probability is not reasonable, for several reasons.

1. Least squares modeling is based on the assumption that y_i is normally distributed (since ε_i is normally distributed). This is obviously not possible when it only takes on two possible values.

2. If a predictor value is large enough or small enough (depending on the sign of the associated slope as long as it is nonzero) a linear model implies that the probability of a success will be outside the range $[0, 1]$, which is impossible.

3. A linear model implies that a specific change in a predictor variable is associated with the same change in the probability of success for any value of the predictor, which is unlikely to be the case. For example, a firm with extremely high debt is likely to already have a high probability of bankruptcy; for such a firm it is reasonable to suppose that an increase in debt of \$1 million (say) would be associated with a small absolute change in the probability of bankruptcy. This is related to point (2), of course, since a probability close to 1 can only increase a small amount and still remain less than or equal to 1. Similarly, for a firm with very little debt (and hence a small probability of bankruptcy), this increase in debt could reasonably be thought to be associated with a small change in the probability of bankruptcy. On the other hand, such an increase in debt could be expected to have a much larger effect on the probability of bankruptcy for a firm with moderate debt, and hence moderate probability of bankruptcy. Similar arguments (but in the opposite direction) would apply to a predictor with an inverse relationship with bankruptcy, such as revenues. This implies "S-shaped" curves for probabilities, as given in Figure 8.1.

In this chapter we discuss **logistic regression**, an alternative regression model that is appropriate for binary data. Many of the inferential methods and techniques discussed in earlier chapters generalize to logistic regression, although a fundamental change is that in the present context they are not exact, but are rather based on approximations.

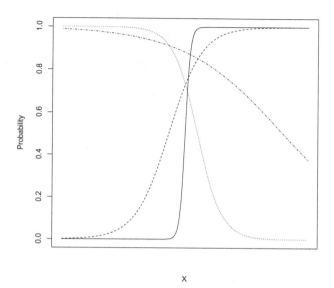

FIGURE 8.1 **S-shaped curves for probabilities.**

8.2   Concepts and Background Material

The regression models discussed thus far are characterized by two key properties: a linear relationship between the expected response and the predictors, and a normal distribution for the errors. Models for a binary response variable are similarly based on two key properties. First, as was pointed out in the previous section, an S-shaped relationship rather than a linear relationship is the basis of the modeling. Second, a distribution that is more appropriate for binary data than the normal distribution is used. We treat each of these points in turn.

8.2.1 THE LOGIT RESPONSE FUNCTION

The functional form underlying logistic regression that generates curves like those in Figure 8.1 is the **logit** function. Let $\pi(\mathbf{x})$ be the probability of a success for the observed values of the predictors $\mathbf{x}$. The *odds* of a success is the ratio of the probability of a success to the probability of failure, or

$$\frac{\pi(\mathbf{x})}{1 - \pi(\mathbf{x})}$$

(since the probability of a failure is $1 - \pi$). Note that the odds vary between 0 and ∞ as the probability varies between 0 and 1.

The logit is defined as the (natural) log of the odds of success,

$$\ell(\mathbf{x}) = \log\left[\frac{\pi(\mathbf{x})}{1 - \pi(\mathbf{x})}\right].$$

Logistic regression hypothesizes that the logit is linearly related to the predictors; that is,

$$\ell(\mathbf{x}) = \log\left[\frac{\pi(\mathbf{x})}{1 - \pi(\mathbf{x})}\right] = \beta_0 + \beta_1 x_{1i} + \cdots + \beta_p x_{pi}. \tag{8.1}$$

Note that the logit varies between $-\infty$ and ∞ as the probability varies between 0 and 1, making logits more suitable for linear fitting. Solving for $\pi(\mathbf{x})$ gives an equivalent representation to (8.1) that explicitly provides the implied S-shaped curve for probabilities,

$$\pi_i(\mathbf{x}) = \frac{e^{\beta_0 + \beta_1 x_{1i} + \cdots + \beta_p x_{pi}}}{1 + e^{\beta_0 + \beta_1 x_{1i} + \cdots + \beta_p x_{pi}}}. \tag{8.2}$$

This inverse of the logit function is sometimes called the **expit** function.

Examination of (8.1) shows that it is, in fact, a semilog model for the odds of success, in the sense of Section 4.3 and equation (4.4). That is, the model posits an additive/multiplicative relationship between a predictor and the odds, with a multiplicative change in the odds of success of e^{β_j} associated with a one unit increase in x_j holding all else in the model fixed. This value is thus called the **odds ratio**, as it represents the ratio of the odds for $x_j + 1$ to the odds for x_j. A slope of 0 corresponds to no relationship between the logit (and hence the probability) and the predictor given the other variables in the model, a positive slope corresponds to a direct relationship, and a negative slope corresponds to an inverse relationship.

The use of natural logs in the definition of the logit implies that if a predictor is going to be modeled in the logged scale, natural logs (base e) should be used, rather than common logs (base 10). This is because in that case the slope β_j is an elasticity (in the sense of Section 4.2). That is, if $\log x_j$ is used as a predictor, this implies that a 1% change in x_j is associated with a β_j% change in the odds of success holding all else in the model fixed.

The logit is not the only function that can generate S-shaped curves. Indeed, any cumulative distribution function for a continuous random variable generates curves similar to those in Figure 8.1, including (for example) the normal distribution (which leads to **probit** regression). The use of the logit response does have some advantages, which will be discussed in Section 8.2.3.

8.2.2 BERNOULLI AND BINOMIAL RANDOM VARIABLES

It is apparent that a normal distribution is not a reasonable choice for data that take on two values, 0 and 1. The natural choice is a Bernoulli random variable, where each observation y_i is independent, with $P(y_i = 1) = \pi_i$.

Combining this with the functional form (8.1) defines the **logistic regression model**.

This model also generalizes in an important way. In some circumstances multiple observations with the same set of predictor variable values occur. This is often by design; for example, a clinical trial might be designed so that exactly 10 men and 10 women each receive a specific dosage of a drug from the set of dosages being studied. In this situation the ith response y_i would correspond to the number of successes (say cures) out of the $n_i = 10$ people of the specific gender who received the specific dosage corresponding to the ith observation. If the responses of each of the 10 individuals are independent of each other, and each individual has the same probability π_i of being cured, then y_i is a binomial random variable based on n_i trials and π_i probability of success [represented $y_i \sim B(n_i, \pi_i)$]. The n_i value is sometimes called the number of replications, as each of the underlying Bernoulli trials has identical values of the predictors. Note that in this situation the number of observations n is the number of response values y_i, not the total number of replications $\sum_i n_i$.

8.2.3 PROSPECTIVE AND RETROSPECTIVE DESIGNS

The particular choice of the logit function to represent the relationship between probabilities and predictors has two main justifications. The first, referred to in Section 8.2.1, is the intuitive nature of (8.1) as a semilog model for the odds of success. Multiplicative relationships are relatively easy to understand, so the interpretation of e^{β_j} as an odds ratio is an appealing one.

The second justification is less straightforward, but of great practical importance. It is related to the sampling design used when obtaining the data. Consider building a model for the probability that a business will go bankrupt as a function of the initial debt carried by the business. There are two ways that we might imagine constructing a sample (say of size 200) of businesses to analyze:

1. Randomly sample 200 businesses from the population of interest. Record the initial debt, and whether or not the business went bankrupt. This is conceptually consistent with following the 200 businesses through time until they either do or do not go bankrupt, and is called a **prospective** sampling scheme for this reason. In the biomedical literature this is often called a **cohort** study. A variation on this design is to sample 100 businesses with low debt and 100 businesses with high debt, respectively; since sampling is based on the value of a predictor (not the response), it is still a prospective design.

2. First consider the set of all businesses in the population that did not go bankrupt; randomly sample 100 of them and record the initial debt. Then consider the set of all businesses in the population that did go bankrupt; randomly sample 100 of them and record the initial debt. This is conceptually consistent with seeing the final state of the businesses first

(bankrupt or not bankrupt), and then going backwards in time to record the initial debt, and is called a **retrospective** sampling scheme for this reason. In the biomedical literature this is often called a **case-control** study.

Note that whether the sampled data are current or from an earlier time has nothing to do with whether a design is prospective or retrospective; the distinction depends only on whether sampling is based on the response outcome (a retrospective study) or not (a prospective study). One way to distinguish between these two sampling approaches is that in a retrospective study the sampling rate is different for successes and for failures; that is, one group is deliberately oversampled while the other is deliberately undersampled so as to get a "reasonable" number of observations in each group.

Each of these sampling schemes has advantages and disadvantages. The prospective study is more consistent with the actual physical process of interest; for example, the observed sample proportion of low-debt businesses that go bankrupt is an estimate of the actual probability of a randomly chosen low-debt business from this population going bankrupt, a number that cannot be estimated using data from a retrospective study (since in that case it was arbitrarily decided that (say) half the sample would be bankrupt businesses, and half would be non-bankrupt businesses). Such studies also have the advantage that they can be used to study multiple outcomes; for example, The British Doctors Study, run by Richard Doll, Austin Bradford Hill, and Richard Peto, followed 40,000 doctors for 50 years, and examined how various factors (particularly smoking) related to different types of cancer, emphysema, heart disease, stroke, and other diseases. On the other hand, if bankruptcy rates are low (say 15%), a sample of size 200 is only going to have about 30 bankrupt businesses in it, which makes it more difficult to accurately model the probability of bankruptcy. A retrospective study can be designed to make sure that there are enough bankrupt companies to estimate their characteristics well.

To simplify things, assume that initial debt is recorded only as Low (L) or High (H). This implies that the data take the form of a 2×2 contingency table (whatever the sampling scheme):

		Bankrupt		
		Yes	*No*	
Debt	*Low*	n_{LY}	n_{LN}	n_L
	High	n_{HY}	n_{HN}	n_H
		n_Y	n_N	n

Here the subscripts Y and N refer to bankrupt ("Yes") and not bankrupt ("No").

Even though the data have the same form, whatever the sampling scheme, the ways these data are generated are very different. The following two tables give the expected counts in the four data cells, depending on the sampling scheme. The π values are conditional probabilities (so, for example, $\pi_{Y|L}$ is the probability of a business going bankrupt given that it has low initial debt):

		PROSPECTIVE SAMPLE					RETROSPECTIVE SAMPLE	
		Bankrupt					**Bankrupt**	
		Yes	*No*				*Yes*	*No*
Debt	*Low*	$n_L \pi_{Y|L}$	$n_L \pi_{N|L}$		**Debt**	*Low*	$n_Y \pi_{L|Y}$	$n_N \pi_{L|N}$
	High	$n_H \pi_{Y|H}$	$n_H \pi_{N|H}$			*High*	$n_Y \pi_{H|Y}$	$n_N \pi_{H|N}$

There is a fundamental difference between the probabilities that can be estimated using the two sampling schemes. For example, what is the probability that a business goes bankrupt given that it has low initial debt? As noted above, this is $\pi_{Y|L}$. It is easily estimated from a prospective sample ($\hat{\pi}_{Y|L} = n_{LY}/n_L$), as can be seen from the left table, but it is impossible to estimate from a retrospective sample. On the other hand, given that a business went bankrupt, what is the probability that it had low initial debt? That is $\pi_{L|Y}$, which is estimable from a retrospective sample ($\hat{\pi}_{L|Y} = n_{LY}/n_Y$), but not from a prospective sample.

The advantage of logistic regression is that the existence of a relationship between debt level and bankruptcy is based on odds ratios (rather than probabilities) through the logit function. In a prospective study, debt is related to bankruptcy only if the odds of bankruptcy versus nonbankruptcy are different for low debt companies than they are for high debt companies; that is, if

$$\frac{\pi_{Y|L}}{\pi_{N|L}} \neq \frac{\pi_{Y|H}}{\pi_{N|H}},$$

or equivalently that the odds ratio does not equal 1,

$$\frac{\pi_{Y|L}\pi_{N|H}}{\pi_{N|L}\pi_{Y|H}} \neq 1.$$

By the definition of conditional probability this odds ratio is equivalent to

$$\frac{\pi_{LY}\pi_{HN}}{\pi_{HY}\pi_{LN}}, \tag{8.3}$$

where (for example) π_{LY} is the probability of a company both having low debt and going bankrupt.

In a retrospective study, in contrast, debt is related to bankruptcy only if the odds of low debt versus high debt are different for bankrupt companies than they are for nonbankrupt companies; that is, if

$$\frac{\pi_{L|Y}}{\pi_{H|Y}} \neq \frac{\pi_{L|N}}{\pi_{H|N}},$$

or equivalently that the odds ratio does not equal 1,

$$\frac{\pi_{L|Y}\pi_{H|N}}{\pi_{L|N}\pi_{H|Y}} \neq 1.$$

Some algebra shows that this odds ratio is identical to (8.3), the one from the prospective study. That is, while the type of conditional probability that can be estimated from the data depends on the sampling scheme, the odds ratio is unambiguous whichever sampling scheme is appropriate. This property generalizes to numerical predictors and multiple predictors. The logit function is the only choice where effects are determined by the odds ratio, so it is the only choice where the measure of the association between the response and a predictor is the same under either sampling scheme. This means that the results of studies based on the same predictors are directly comparable to each other even if some are based on prospective designs while others are based on retrospective designs.

Since odds ratios are uniquely defined for both prospective and retrospective studies, the slope coefficients $\{\beta_1, \beta_2, \ldots, \beta_p\}$ are also uniquely defined. The constant term β_0, however, is driven by the observed proportions of successes and failures, so it is affected by the construction of the study. Thus, as noted above, in a retrospective study, the results of a logistic regression fit cannot be used to estimate the (prospective) probability of success, since that depends on a correct estimate of β_0 through (8.2). Let π_Y and π_N be the true (unconditional) probabilities that a randomly chosen business goes bankrupt or does not go bankrupt, respectively. These numbers are called **prior probabilities**. If prior probabilities of success (π_Y) and failure (π_N) are available, the constant term in a fitted logistic regression can be adjusted so that correct (prospective) probabilities can be estimated. The adjusted constant term has the form

$$\tilde{\beta}_0 = \hat{\beta}_0 + \log\left(\frac{\pi_Y n_N}{\pi_N n_Y}\right)$$

(the method used to determine $\hat{\beta}_0$ is discussed in the next section). Since often a data analyst is not sure about exactly what (π_Y, π_N) are, it is reasonable to try a range of values to assess the sensitivity of the estimated probabilities based on the adjusted intercept to the specific choice that is made.

8.3 Methodology

8.3.1 MAXIMUM LIKELIHOOD ESTIMATION

As was noted earlier, least squares estimation of the logistic regression parameters β is not the best choice, as that is appropriate for normally distributed data. The generalization of least squares to binomially-distributed data is **maximum likelihood estimation**. The theory of maximum likelihood is beyond the scope of this book, but the underlying principle is that parameters are estimated with the values that give the observed data the maximum probability of occurring. For binary/binomial data, this corresponds to maximizing the log-likelihood function

$$L = \sum_{i=1}^{n} [y_i \log \pi_i + (n_i - y_i) \log(1 - \pi_i)], \qquad (8.4)$$

where π_i is assumed to satisfy the expit relationship (8.2) with the predictors (terms that are not functions of the parameters are omitted from L since they do not affect estimation). A related value is the **deviance**, which is twice the difference between the maximum possible value of the log-likelihood L and the value for the fitted model. A small value of the deviance means that the observed data have almost as high a probability of occurring based on the fitted model as is possible. It can be shown that maximum likelihood estimates possess various large-sample (asymptotic) optimality properties.

Least squares corresponds to maximum likelihood for errors that are normally distributed, but in general (and for logistic regression in particular) maximum likelihood estimates cannot be determined in closed form, but rather are obtained via an iterative algorithm. One such algorithm (iteratively reweighted least squares, or IRWLS) shows that the maximum likelihood estimates are approximately weighted least squares estimates (6.5), with a weight and corresponding error variance for each observation that depends on the parameter estimates. This approximation is the basis of many of the inferential tools used to assess logistic regression fits. Chapter 3 of Hilbe (2009) provides more details about this algorithm.

8.3.2 INFERENCE, MODEL COMPARISON, AND MODEL SELECTION

Inferential questions arise in logistic regression that are analogous to those in least squares regression, but the solutions are more complicated, as they are either highly computationally intensive or are based on approximations. For example, a test of the overall strength of the regression, testing the hypotheses

$$H_0 : \beta_1 = \cdots = \beta_p = 0$$

versus

$$H_a : \beta_j \neq 0 \text{ for at least one } j$$

(which for least squares fitting is tested using the overall F-test) is desirable. The standard test of these hypotheses is the **likelihood ratio test**, which is based on comparing the strength of the fit without any predictors to the strength of the fit using predictors, as measured by the difference in the deviance values for the models with and without predictors. The likelihood ratio test for the overall significance of the regression is

$$LR = 2 \sum_{i=1}^{n} \left[y_i \log \left(\frac{\hat{\pi}_i^a}{\hat{\pi}_i^0} \right) + (n_i - y_i) \log \left(\frac{1 - \hat{\pi}_i^a}{1 - \hat{\pi}_i^0} \right) \right], \qquad (8.5)$$

where $\hat{\pi}^a$ are the estimated probabilities based on the fitted logistic regression model, and $\hat{\pi}^0$ are the estimated probabilities under the null hypothesis. This is compared to a χ^2 distribution on p degrees of freedom, which is valid as long as either n is large or the n_i values are reasonably large.

Two tests of the additional predictive power provided by an individual predictor given the others are also available. These are tests of the hypotheses

$$H_0 : \beta_j = 0$$

versus

$$H_a : \beta_j \neq 0.$$

These hypotheses can be tested using the likelihood ratio test form described above, where $\hat{\pi}^0$ is calculated based on all predictors except the jth predictor and $\hat{\pi}^a$ is calculated based on all predictors, which is then compared to a χ^2 distribution on 1 degree of freedom. This requires fitting $p + 1$ different models to test the significance of the p slopes, and is therefore not the typical approach. The alternative (and standard) approach is to use the so-called **Wald test statistic,**

$$z_j = \frac{\hat{\beta}_j}{\widehat{s.e.}(\hat{\beta}_j)},$$

where $\widehat{s.e.}(\hat{\beta}_j)$ is calculated based on the IRWLS approximation to the maximum likelihood estimate. This is analogous to a t-test for least squares fitting, but is again based on asymptotic assumptions, so the statistics are compared to a Gaussian distribution to determine significance rather than a t-distribution. Confidence intervals for individual coefficients take the form $\hat{\beta}_j \pm z_{\alpha/2}\widehat{s.e.}(\hat{\beta}_j)$, and a confidence interval for the associated odds ratio can be obtained by exponentiating each of the endpoints of the confidence interval for β_j.

Any two models where one is a special case of the other based on a linear restriction can be compared using hypothesis testing. The simpler model represents the null while the more complicated model represents the alternative, and they are tested using the likelihood ratio test as a difference of deviance values. The appropriate degrees of freedom for the χ^2 approximation is the difference in the number of parameters estimated in the two models. This is analogous to the partial F-test in least squares regression.

Such tests can be useful tools for model comparison and selection, but as was discussed in Section 2.3.1, hypothesis tests are not necessarily the best tools for model selection. The AIC criterion is applicable in logistic regression models, taking the form

$$AIC = -2L + 2(p + 1),$$

where L is the log-likelihood (8.4). Equivalently $-2L$ can be replaced with LR from (8.5) when calculating AIC. Although the theory underlying AIC_c is not applicable for logistic regression, practical experience suggests that it

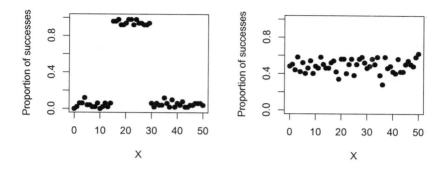

FIGURE 8.2 Two plots of hypothetical observed proportions of success versus a predictor.

can help guard against the tendency of AIC to choose models that are overly complex, so it can still be a useful tool. As before, it has the form

$$AIC_c = AIC + \frac{2(p+2)(p+3)}{n-p-3}.$$

8.3.3 GOODNESS-OF-FIT

The logistic model (8.1) or (8.2) is a reasonable one for probabilities, but may not be appropriate for a particular data set. This is not the same thing as saying that the predicting variables are not good predictors for the probability of success. Consider the two plots in Figure 8.2. The variable X is a potential predictor for the probability of success, while the vertical axis gives the observed proportion of successes in samples taken at those values of X. So, for example, X could be the dosage of a particular drug, and the target variable is the proportion of people in a trial that were cured when given that dosage.

In the plot on the left, X is very useful for predicting success, but the linear logistic regression model does not fit the data, since the probability of success is not a monotone function of X (in a situation like this treating X as categorical, with the three categories being $\{[0, 15), [15, 30), [30, 50]\}$, would seem to be a much more sensible strategy). In the plot on the right, X is not a useful predictor for success (the probability of success appears to be unrelated to X), but the linear logistic regression model fits the data, as a very flat S-shaped curve (with $\beta_X \approx 0$) goes through the observed proportions of success reasonably well. **Goodness-of-fit tests** are designed to assess the fit of a model through the use of hypothesis testing. Such statistics test the hypotheses

H_0 : The linear logistic regression model fits the data

versus

H_a : The linear logistic regression model does not fit the data.

Such tests proceed by comparing the variability of the observed data around the fitted model to the data's inherent variability. This is possible because for a binomial random variable the observed model-based variability (based on the residual $y_i - n_i\hat{\pi}_i$) is distinct from the inherent variability of the random variable [based on the fact that $V(y_i) = n_i\pi_i(1 - \pi_i)$].

There are two standard goodness-of-fit test statistics. The **Pearson goodness-of-fit statistic** equals

$$X^2 = \sum_{i=1}^{n} \frac{(y_i - n_i\hat{\pi}_i)^2}{n_i\hat{\pi}_i(1 - \hat{\pi}_i)}, \tag{8.6}$$

while the **deviance statistic** (mentioned earlier) equals

$$G^2 = 2\sum_{i=1}^{n} \left[y_i \log\left(\frac{y_i}{n_i\hat{\pi}_i}\right) + (n_i - y_i) \log\left(\frac{n_i - y_i}{n_i(1 - \hat{\pi}_i)}\right) \right] \tag{8.7}$$

(the latter statistic is sometimes referred to as the *residual deviance*). When the n_i values are reasonably large ($n_i > 5$, with some values perhaps even smaller), each of these statistics is referenced to a χ^2 distribution on $n - p - 1$ degrees of freedom under the null hypothesis that the logistic regression model fits the data. Thus, a small tail probability suggests that the linear logistic regression model is not appropriate for the data, and an alternative should be sought. The signed square roots of the values on the righthand side of (8.6) are called the **Pearson residuals**. It should be noted that care must be taken to account for the fact that the form (and implication) of these tests can be different when using different statistical software when $n_i > 1$, as they depend on how the software defines the observations i (whether responses are defined based on the binomial responses with n_i replications for each observation, as is done here, or based on the underlying Bernoulli 0/1 outcomes). See Simonoff (1998a) for fuller discussion of this point.

Unfortunately, these tests are not trustworthy when the n_i values are small, and are completely useless in the situation of Bernoulli response data [$n_i = 1$, so the response for each observation is simply success ($y = 1$) or failure ($y = 0$)]. This is the justification for a third goodness-of-fit test, the **Hosmer-Lemeshow test**.

This test is constructed based on a Pearson goodness-of-fit test, but one where observations are grouped together in a data-dependent way to form rough replications. First, all of the observations are ordered by their estimated success probabilities $\hat{\pi}$. The observations are then divided into g roughly equisized groups, with g usually taken to be 10 except when that would lead to too few observations in each group. Treating this new categorization of the data as if it was the original form of the data, with the g groups defining

g observations with replications, implies observed and expected numbers of successes for each group. The Hosmer-Lemeshow test uses these by calculating the usual X^2 test based on the new categorization, which is compared to a χ^2 distribution on $g - 2$ degrees of freedom. It should be noted, however, that even the Hosmer-Lemeshow test is suspect when the expected numbers of successes or failures in the constructed categorization are too small (less than two or three, say). Other alternative statistics for the $n_i = 1$ situation have also been proposed.

8.3.4 MEASURES OF ASSOCIATION AND CLASSIFICATION ACCURACY

While tests of hypotheses are useful to assess the strength of a logistic regression relationship, they do not address the question of whether the relationship is of practical importance, as (for example) R^2 can for least squares regression. Several measures of association have been proposed for this purpose, which are closely related to each other. Start with a fitted logistic regression model with resultant fitted probabilities of success for each of the observations. Consider each of the pairs (i, j) of observations where one observation is a success (i) and the other is a failure (j). Each of these has a corresponding pair $(\hat{\pi}_i, \hat{\pi}_j)$. A "good" model would have a higher estimated probability of success for the observation that was actually a success than for the observation that was actually a failure; that is, $\hat{\pi}_i > \hat{\pi}_j$. Such a pair is called *concordant*. If for a given pair $\hat{\pi}_i < \hat{\pi}_j$, the pair is called *discordant*. A model that separates the successes from the failures well would have a high proportion of concordant pairs and low proportion of discordant ones. There are no formal cutoffs for what constitutes a "good enough" performance here, but observed values can be compared for different possible models to assess relative performance in this sense.

Various statistics are based on the observed proportions of concordant and discordant pairs. **Somers' D**, for example, is the difference between the proportions of concordant and discordant pairs. Somers' D is equivalent to the well-known area under the Receiver Operating Characteristic (ROC) curve ($AUR = D/2 + .5$), and also the Wilcoxon-Mann-Whitney rank sum test statistic for comparing the distributions of probability estimates of observations that are successes to those that are failures [$WMW = AUR \times n_S n_F$, where n_S (n_F) is the number of successes (failures)]. Note that while each of these is a measure of the quality of the probability rankings implied by the model (in the sense of concordance), a good probability ranking need not necessarily be *well-calibrated*. For example, if the estimated probability of success for each observation was exactly one-half the true probability, the probability rankings would be perfect (implying $D = AUR = 1$), but not well-calibrated, since the estimates were far from the true probabilities.

In the medical diagnostic testing literature, the following rough guide for interpretation of D has been suggested (Hosmer and Lemeshow, 2000, p. 162, provides similar guidelines). It is perhaps useful as a way to get a sense of what the value is implying, but should be recognized as being fairly arbitrary, and should not be taken overly seriously.

Range of D	Rough interpretation
$0.8 - 1.0$	Excellent separation
$0.6 - 0.8$	Good separation
$0.4 - 0.6$	Fair separation
$0.2 - 0.4$	Poor separation
$0.0 - 0.2$	Little to no separation

Logistic regression also can be used for prediction of group membership, or *classification*. This is typically appropriate when operating at the 0/1 (Bernoulli) response level. After estimating β, (8.2) can be used to give an estimate of the probability of success for that observation. A success/failure prediction for an observation is obtained based on whether the estimated probability is greater than or less than a cutoff value. This value is often taken to be .5, although in some situations another choice of cutoff might be preferable (based, for example, on the relative costs of misclassifying a success as a failure and vice versa). If this process is applied to the original data that was used to fit the logistic regression model, a **classification table** results. The resultant table takes this form:

Predicted result

		Success	Failure	
Actual	Success	n_{SS}	n_{SF}	$n_{S.}$
result	Failure	n_{FS}	n_{FF}	$n_{F.}$
		$n_{.S}$	$n_{.F}$	n

The proportion of observations correctly classified is $(n_{SS} + n_{FF})/n$, and the question is then whether this is a large number or not. The answer to this question is not straightforward, because the same data were used to both build the model and evaluate its ability to do classifications (that is, we have used the data twice). As a result the observed proportion correctly classified can be expected to be biased upwards compared to the situation where the model is applied to completely new data.

The best solution to this problem is the same as it is when evaluating the predictive performance of least squares models — validate the model on new data to see how well it classifies new observations. In the absence of new data, two diagnostics have been suggested that can be helpful. A lower bound for what could be considered reasonable performance is the **base rate**, which is the proportion of the sample that comes from the larger group (sometimes termed $C_{\max}$ in this context).

A more nuanced argument is as follows. If the logistic regression had no power to make predictions, the actual result would be independent of the predicted result. That is, for example,

P(Actual result a success and Predicted result a success) =
$$P(\text{Actual result a success}) \times P(\text{Predicted result a success}).$$

The right side of this equation can be estimated using the marginal probabilities from the classification table, yielding

$$P(\text{Actual result a success and Predicted result a success}) = \left(\frac{n_{S\cdot}}{n}\right)\left(\frac{n_{\cdot S}}{n}\right).$$

A similar calculation for the failures gives

$$P(\text{Actual result a failure and Predicted result a failure}) = \left(\frac{n_{F\cdot}}{n}\right)\left(\frac{n_{\cdot F}}{n}\right).$$

The sum of these two numbers is an estimate of the expected proportion of observations correctly classified if the actual and predicted memberships were independent, so achieving this level of classification accuracy would not be evidence of a useful ability to classify observations. Since a higher observed correct classification proportion is expected because the data were used twice, this number is typically inflated by 25% before being compared to the observed correct classification proportion, resulting the so-called C_{pro} measure.

8.3.5 DIAGNOSTICS

Unusual observations can have as strong an effect on a fitted logistic regression as in linear regression, and therefore need to be identified and explored. The IRWLS representation of the maximum likelihood logistic regression estimates provides the mechanism to construct approximate versions of diagnostics such as standardized (Pearson) residuals, leverage values, and Cook's distances, using the implied hat matrix and observation variances from IRWLS as if the model is a WLS fit, as is described in Section 6.3.3.

8.4 Example — Smoking and Mortality

In 1972–1974 a survey was taken in Whickham, a mixed urban and rural district near Newcastle upon Tyne, United Kingdom (Appleton et al., 1996). Twenty years later a followup study was conducted, and it was determined if the interviewee was still alive. Among the information obtained originally was whether a person was a smoker or not and their age, divided into seven categories. The data can be summarized in the following table:

Age group	Smoking status	Survived	At risk
18–24	Smoker	53	55
18–24	Nonsmoker	61	62
25–34	Smoker	121	124
25–34	Nonsmoker	152	157
35–44	Smoker	95	109
35–44	Nonsmoker	114	121
45–54	Smoker	103	130
45–54	Nonsmoker	66	78
55–64	Smoker	64	115
55–64	Nonsmoker	81	121
65–74	Smoker	7	36
65–74	Nonsmoker	28	129
75 and older	Smoker	0	13
75 and older	Nonsmoker	0	64

As always a good first step in the analysis is to look at the data. In this case there are two predictors, age and smoking status. A simple cross-classification shows that twenty years later 76.1% of the 582 smokers were still alive, while only 68.6% of 732 nonsmokers were still alive. That is, smokers had a higher survival rate than nonsmokers, a pattern that seems puzzling at first glance.

Figure 8.3 gives more reasonable representations of the data. Since there are multiple interviewees in each of the age categories for both smokers and nonsmokers, the observed proportions of people who were alive 20 years later

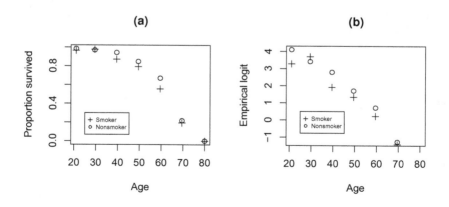

FIGURE 8.3 Plots for the Whickham smoking and mortality data. (a) Empirical survival proportions, separated by age group and smoking status. (b) Empirical logits, separated by age group and smoking status.

are reasonable estimates of the true underlying probabilities. The plots treat age as a numerical variable taking on a value at the midpoint of the interval and using the value 80 for the last age group. The left plot shows the observed survival proportions $\bar{\pi}_i$ versus age, while the right plot gives the empirical logits $\log[\bar{\pi}_i/(1 - \bar{\pi}_i)]$. Note that the empirical logit is not defined for the "75 or older" age group since no interviewees from that group were alive at followup. If a constant shift linear logistic regression model is reasonable, we would expect to see two roughly parallel linear relationships in the right plot, one for smokers and one for nonsmokers. This seems to be reasonable for younger ages, but the empirical logits are too low for the 65–74 group (and would be $-\infty$ for the 75 and older group), suggesting a possible violation of linearity in the logit scale.

It is also apparent that for most age groups survival is lower for smokers than nonsmokers, as would be expected. This reversal of direction from the marginal relationship (higher survival rates for smokers than for nonsmokers) to the conditional one (lower survival rates for smokers than for nonsmokers given age) is familiar in any multiple regression situation (recall that it can lead to the misconception that a multiple regression slope coefficient has the "wrong sign" discussed in Section 1.3.1), and in the context of categorical data is called **Simpson's paradox**. The underlying reason, of course, is correlation between predictors, which in this case corresponds to higher smoking rates for younger interviewees than for older ones (49.7% among interviewees less than 65 years old versus 20.2% for those at least 65 years old). As a result what appears to be a positive effect of smoking (higher survival rates for smokers) is actually the positive effect of being younger at the time of original interview.

Output for a fitted logistic regression is as follows:

```
Coefficients:
            Estimate Std. Error z value Pr(>|z|)
(Intercept)  7.687751   0.447646   17.174   <2e-16 ***
Age         -0.124957   0.007274  -17.178   <2e-16 ***
Smoker      -0.266053   0.168702   -1.577    0.115
---
Signif. codes:
  0 '***' 0.001 '**' 0.01 '*' 0.05 '.' 0.1 ' ' 1

    Null deviance: 641.496  on 13  degrees of freedom
Residual deviance:  32.572  on 11  degrees of freedom
AIC: 85.568
```

The overall regression is highly statistically significant ($LR = 608.9$ on 2 degrees of freedom, $p \approx 0$, obtained as the difference between the null and residual deviances). The Wald test for age is also highly statistically significant, but that for smoking status is only marginal. Exponentiating the slopes gives odds ratios of 0.883 and 0.766, respectively, implying that (given smoking status) being an additional year older is associated with an estimated 11.7% smaller odds of being alive 20 years later, and given age being a smoker is

TABLE 8.1 Details of models fit to the Whickham smoking study data.

Model	LR	G^2 (p-value)	X^2 (p-value)	AIC
Linear	608.9	32.6 (< .001)	29.3 (.002)	85.6
Quadratic	629.9	11.6 (.311)	9.6 (.474)	66.6
Quadratic versus Linear: $LR = 20.9$, $df = 1$, $p < .001$				
Categorical	639.1	2.4 (.882)	2.4 (.882)	65.4
Categorical versus Linear: $LR = 30.2$, $df = 5$, $p < .001$				
Categorical versus Quadratic: $LR = 9.2$, $df = 4$, $p = .055$				

associated with an estimated 23.4% lower odds of survival (the constant odds ratio for smoking for all ages reflects the constant shift nature of the fitted model).

Unfortunately, goodness-of-fit tests imply that this model does not fit very well, as $G^2 = 32.6$ ($p < .001$) and $X^2 = 29.3$ ($p = .002$). Figure 8.3(b) provides a clue as to why, since (as was noted earlier) the relationship between age and the empirical logits is apparently not linear. The figure suggests (at least) two possible ways of enriching the model while still maintaining a constant shift for smoking status: including a quadratic function of age, or treating age as a categorical variable (since it is actually only given as membership in one of seven categories). Table 8.1 summarizes these fits, and how they compare to the original (constant shift) linear model.

The quadratic and categorical models are both clear improvements over the linear model, with much lower AIC values and highly statistically significant LR tests comparing them (the linear model is a special case of the quadratic model, and both are special cases of the categorical model). The choice between the quadratic and categorical models is less obvious; while the categorical model has lower AIC, the difference is small, and the hypothesis that the quadratic model is adequate compared to the categorical model is only weakly rejected ($p = .055$). Fortunately, from the point of view of the relationship between smoking and mortality, the choice is moot; in either case the estimated slope for the smoking indicator is roughly -0.43, implying 35% lower odds of survival twenty years later given age.

Output for the model treating age as a categorical variable is given below.

```
Coefficients:
                       Estimate Std. Error z value Pr(>|z|)
(Intercept)              3.8601     0.5939   6.500 8.05e-11 ***
Age group = 29.5        -0.1201     0.6865  -0.175 0.861178
Age group = 39.5        -1.3411     0.6286  -2.134 0.032874 *
Age group = 49.5        -2.1134     0.6121  -3.453 0.000555 ***
Age group = 59.5        -3.1808     0.6006  -5.296 1.18e-07 ***
Age group = 69.5        -5.0880     0.6195  -8.213  < 2e-16 ***
```

```
Age group = 80      -27.8073 11293.1437  -0.002 0.998035
Smoker               -0.4274      0.1770  -2.414 0.015762 *
---
Signif. codes:
  0 '***' 0.001 '**' 0.01 '*' 0.05 '.' 0.1 ' ' 1

   Null deviance: 641.4963  on 13  degrees of freedom
Residual deviance:   2.3809  on  6  degrees of freedom
AIC: 65.377
```

The age group variable is fit using six indicator variables, taking the youngest category (18-24 years old) as the reference group. A striking result in the output are the strange entries for the `Age group = 80` variable; the estimated slope is extremely large and negative, but the standard error is so large that the p-value for the Wald test of whether the slope equals 0 is virtually 1. The reason for this is that none of the interviewees in the 75 and older group were alive at the time of followup, so the model is trying to estimate the probability of survival of that group as 0. Based on (8.2) that can only occur for a slope of $-\infty$, and for that reason the software's iterative algorithm pushes the estimated slope to be as negative as possible. A similar pattern would occur if everyone in an age group had survived, only then the coefficient would be extremely large and positive rather than negative.

8.5 Example — Modeling Bankruptcy

As was stated at the beginning of this chapter, the study of bankruptcy of companies has direct parallels to the study of survival of people. The data discussed here were presented in Section 9.2 of Simonoff (2003), and are based on a retrospective sample of 25 telecommunications firms that declared bankruptcy between May 2000 and January 2002 that had issued financial statements for at least two years, and information from the December 2000 financial statements of 25 telecommunications that did not declare bankruptcy. Five financial ratios (each expressed as a percentage) were chosen as potential predictors of bankruptcy, details of which can be found in Simonoff (2003):

1. Working capital as a percentage of total assets (WC/TA), a measure of liquidity.

2. Retained earnings as a percentage of total assets (RE/TA), a measure of cumulative profitability over time.

3. Earnings before interest and taxes as a percentage of total assets (EBIT/TA), a measure of the productivity of a firm's assets.

4. Sales as a percentage of total assets (S/TA), a measure of the ability of a firm's assets to generate sales.

5. Book value of equity divided by book value of total liabilities (BVE/BVL), a measure of financial leverage (that is, debt).

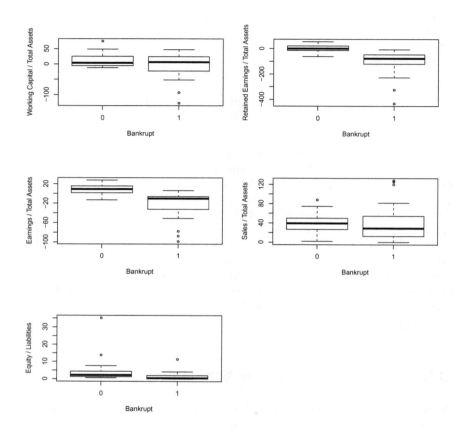

FIGURE 8.4 Side-by-side boxplots for the bankruptcy data.

In contrast to the Whickham smoking study data, these data take the form of 50 observations, each with a 0/1 response variable, so it is not possible to look at the data by plotting empirical proportions (or logits) versus predictors. A good alternative is to construct side-by-side boxplots for each predictor, separating on the response category. This doesn't necessarily imply that a linear logistic model is appropriate, but is still helpful in seeing if a predictor discriminates between successes and failures.

Figure 8.4 gives these side-by-side boxplots. Working capital, retained earnings, and earnings before interest and taxes each show strong separation between bankrupt and nonbankrupt firms, in the ways that would have been expected. Note that while (as always) there is no assumption regarding the distributions of the predictors, the long right tail in the equity variable suggests that logging this variable (using natural logs) might be helpful. In fact, there were no appreciable differences in the implications of the analysis when either logged or unlogged versions are used, so the unlogged variable is used here.

As a first step, a logistic regression model can be fit based on all of the predictors. This results in the following output:

```
Coefficients:
              Estimate  Std. Error  z value  Pr(>|z|)
(Intercept)    7.42646     6.35770    1.168     0.243
WC.TA         -0.15587     0.12208   -1.277     0.202
RE.TA         -0.07605     0.06311   -1.205     0.228
EBIT.TA       -0.49111     0.32260   -1.522     0.128
S.TA          -0.08040     0.09216   -0.872     0.383
BVE.BVL       -2.07764     1.47488   -1.409     0.159


    Null deviance: 69.315  on 49  degrees of freedom
Residual deviance: 11.847  on 44  degrees of freedom
AIC: 23.847
```

While the overall regression is highly statistically significant ($LR = 57.5$ on 5 degrees of freedom, $p < .001$), all of the individual Wald tests have moderately high p-values. This suggests the possibility of simplifying the model, but an index plot of the standardized residuals (Figure 8.5) shows that the first observation is a clear outlier (this observation shows up as an outlier in any reasonable simplified model as well). This is the firm 360Networks. It was one of only two firms that ultimately went bankrupt that had positive earnings the year before insolvency, and had \$6.3 billion in total assets three months before it declared bankruptcy because of thousands of miles of cable that it owned.

A logistic regression model based on all of the predictors after omitting this observation results in the following output:

```
Coefficients:
             Estimate  Std. Error   z value  Pr(>|z|)
(Intercept)   210.732   48078.738     0.004     0.997
WC.TA          -3.415    1032.286    -0.003     0.997
RE.TA          -1.206     424.416    -0.003     0.998
EBIT.TA       -13.553    2960.215    -0.005     0.996
S.TA           -2.265     618.155    -0.004     0.997
BVE.BVL       -61.575   15401.227    -0.004     0.997


    Null deviance: 6.7908e+01  on 48  degrees of freedom
Residual deviance: 5.2176e-08  on 43  degrees of freedom
AIC: 12
```

This model fits the data perfectly, as the deviance is 0. This is termed **complete separation** (since the predictors completely separate the bankrupt firms from the nonbankrupt ones), and results in all of the estimated standard errors of the estimated slopes being extremely large (and the Wald statistics being correspondingly deflated). Recall that a similar problem arose in the

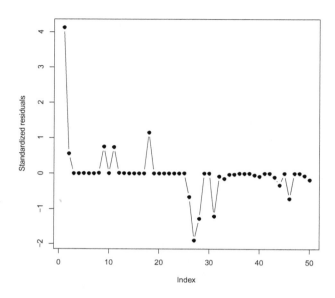

FIGURE 8.5 **Index plot of standardized (Pearson) residuals for the logistic regression fit to the bankruptcy data.**

Whickham smoking survey data (page 166) for the same reason; the only way for a logistic regression to estimate a probability as 0 or 1 is to send the slope coefficient to $\pm\infty$.

A viable solution to this problem is to find a simpler model, if possible, that fits almost as well, but where the maximum likelihood estimates are finite. Table 8.2 summarizes the best models for each number of predictors p. All of the models are highly statistically significant, with LR statistics at least 43.7, so these are not reported in the table. All of the models fit well according to the Hosmer-Lemeshow test. The best model according to AIC and AIC_c is the four-predictor model that omits S/TA, which has a perfect fit. The three-predictor model based on RE/TA, EBIT/TA, and BVE/BVL, however, fits almost perfectly (with Hosmer-Lemeshow statistic $H\text{-}L = 0.6$ and Somers' $D = 0.99$), and has finite (and hence interpretable) estimated slope coefficients. Note that neither the Pearson goodness-of-fit test X^2 or the deviance G^2 should be examined for these data, since the observations have no replications ($n_i = 1$ for all i).

Output for the three-predictor model is as follows.

TABLE 8.2 Details of models fit to the bankruptcy data with 360Networks omitted. *H-L* refers to the Hosmer-Lemeshow goodness-of-fit test, and D refers to Somers' D.

p	WC/ TA	RE/ TA	EBIT /TA	S/ TA	BVE/ BVL	$H - L$ (*p*-value)	D	AIC	AIC_c
0							0.00	69.9	70.2
1		X				5.5 (.707)	0.93	28.3	28.8
2		X			X	4.8 (.778)	0.98	20.9	21.8
3		X	X		X	0.6 (> .999)	0.99	17.4	18.8
4	X	X	X		X	0.0 (1.000)	1.00	10.0	12.0
5	X	X	X	X	X	0.0 (1.000)	1.00	12.0	14.7

```
Coefficients:
              Estimate Std. Error  z value  Pr(>|z|)
(Intercept)  -0.09166    1.47135   -0.062    0.9503
RE.TA        -0.08229    0.04230   -1.945    0.0517 .
EBIT.TA      -0.26783    0.15854   -1.689    0.0912 .
BVE.BVL      -1.21810    0.76536   -1.592    0.1115
---
Signif. codes:
   0 '***' 0.001 '**' 0.01 '*' 0.05 '.' 0.1 ' ' 1

    Null deviance: 67.9080  on 48  degrees of freedom
Residual deviance:  9.3841  on 45  degrees of freedom
AIC: 17.384
```

The Wald statistics are at first glance surprisingly small for a model with such strong fit, but this is actually not uncommon. The estimated standard error used in the denominator of the statistic is known to become too large when an alternative hypothesis far from the null is actually true because of the use of parameter estimates rather than null values (Mantel, 1987), which deflates the statistic. The likelihood ratio test for the significance of each variable does not suffer from this difficulty, and each of these (not reported here) is highly statistically significant. All three coefficients are negative, as would be expected. Exponentiating the slopes gives odds ratios, given the other variables are held fixed; for example, $e^{-.082} = .92$, implying that a one percentage point increase in RE/TA is associated with an estimated 8% decrease in the odds of a firm going bankrupt, given EBIT/TA and BVE/BVL are held fixed. Regression diagnostics do not indicate any outliers, although there are two leverage points, IDT Corporation and eGlobe, the former of which also has a relatively large Cook's distance (see Figure 8.6). Given the strength of the regression, it is not surprising that omitting these observations does not change the implications of the modeling in a fundamental way.

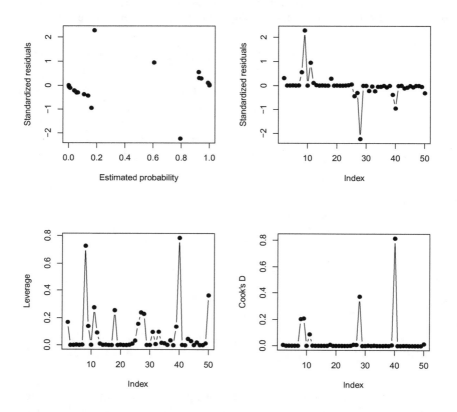

FIGURE 8.6 Diagnostic plots for the three-predictor model fit to the bankruptcy data with 360Networks omitted.

A classification table for this fitting is given below. Forty-seven of 49, or 95.9%, of the firms were correctly classified, far higher than

$$C_{\text{pro}} = (1.25)[(.4898)(.4898) + (.5102)(.5102)] = 62.5\%$$

and $C_{\text{max}} = 51.0\%$, reinforcing the strength of the logistic regression.

		Predicted result		
		Bankrupt	Not bankrupt	
Actual	Bankrupt	23	1	24
result	Not bankrupt	1	24	25
		24	25	

Since this is a retrospective study, the estimated probabilities of bankruptcy are not appropriate for prospective modeling, but prospective probabilities can be obtained by adjusting the constant term of the regression using prior probabilities of bankruptcy. Say a 10% bankruptcy prior probability is used, which is roughly consistent with what would be expected for firms with a corporate bond rating of B. This then yields the adjusted intercept

$$\tilde{\beta}_0 = \hat{\beta}_0 + \log\left[\frac{(.10)(25)}{(.90)(24)}\right] = \hat{\beta}_0 - 2.1564 = -2.2481,$$

which would be used in (8.2).

8.6 Summary

In this chapter we have discussed the application of logistic regression to the modeling of binary response data. Hosmer and Lemeshow (2000) provides a much more detailed discussion of logistic regression, and the discussion in Chapter 9 of Simonoff (2003) ties this material more closely to the broader analysis of categorical data.

In recent years increased computing power has allowed for the possibility of replacing the standard asymptotic inference tools with exact methods. Such tests and estimates are based on a conditional analysis, where inference proceeds conditionally on the sufficient statistics for parameters not of direct interest. Rather than appealing to approximate normal and χ^2 distributions, permutation distributions from the conditional likelihood are used, which can require a good deal of computing power and special software. Simonoff (2003) provides further discussion of these methods for categorical data inference.

Logistic regression is of course based on the assumption that the response y_i follows a binomial distribution. This is potentially problematic for data where there are replications for the ith observation (that is, $n_i > 1$). In that case the binomial assumption can be violated if there is correlation among the n_i individual responses, or if there is heterogeneity in the success probabilities that hasn't been modeled. Each of these violations can lead to **overdispersion**, where the variability of the probability estimates is larger than would be implied by a binomial random variable. In this situation methods that correct for the overdispersion should be used, either implicitly (such as through what is known as **quasi-likelihood**) or explicitly (fitting a regression model using a response distribution that incorporates overdispersion, such as the beta-binomial distribution). Simonoff (2003) provides further discussion, and the corresponding problem for count data regression will be discussed in more detail in Section 10.4.

Logistic regression for binomial data and (generalized) least squares regression for Gaussian data are both special cases of the family of **generalized linear models**. Such models also have direct applicability in the regression analysis of count data, and will be discussed further in Chapter 10.

KEY TERMS

Area under the ROC curve: A measure of the ability of a model to separate successes from failures. It is (literally) the area under the ROC curve, and ranges from 0.5) (no separation) to 1.0 (perfect separation). It is equivalent to Somers' D through the relationship $AUR = D/2 + .5$.

Case-control design: A study in which sampling of units is based on the response categories that are being studied. This is also referred to as a retrospective sampling scheme.

Cohort design: A study in which sampling of units is not based on the outcome responses, but rather either randomly from the population or perhaps based on predictor(s). Such studies sometimes follow units over long periods of time, observing outcome(s) as they manifest themselves. This is also referred to as a prospective sampling scheme.

Complete separation: The condition when a classification procedure correctly identifies all of the observations as successes or failures. This can indicate a model that is overspecified, since a simpler model that correctly classifies the vast majority (but not all) of the observations is often a better representation of the underlying process.

Deviance: A measure of the fit of a model fitted by the maximum likelihood method. The deviance equals twice the difference between the log-likelihood for the saturated model (the model with maximum log-likelihood) and that of the fitted model. Under certain conditions it has an asymptotic χ^2 distribution with appropriate degrees of freedom and can be used as a measure of goodness-of-fit, but is not applicable for this purpose when the number of replications per observation is small.

Expit function: The inverse of the logit function giving the relationship between the probability π and the logit ℓ, $\pi = \exp(\ell)/[\exp(\ell) + 1]$.

Goodness-of-fit statistic: A hypothesis test of whether an observed model provides an adequate fit to the data.

Hosmer-Lemeshow goodness-of-fit statistic: A goodness-of-fit test statistic designed for the situation when the number of replications for (some) observations is small.

Likelihood ratio test: A test in which a ratio of two likelihoods (or more accurately the difference between two log-likelihoods) is used to judge the validity of a statistical hypothesis. Such tests can be used to test a wide range of statistical hypotheses, often analogous to those based on F-tests in least squares regression. Under certain conditions it has an asymptotic χ^2 distribution with appropriate degrees of freedom.

Logistic regression: A regression model defining a (typically linear) relationship between the logit of the response and a set of predictors.

Logit function: The inverse of the expit function giving the relationship between the logit ℓ and the probability π, $\ell = \log[\pi/(1-\pi)]$ (that is, the log of the odds).

Maximum likelihood estimation: A method of estimating the parameters of a model in which the estimated parameters maximize the likelihood for the data. Least squares estimation is a special case for Gaussian data. Maximum likelihood estimates are known to possess various asymptotic optimality properties.

Odds ratio: The multiplicative change in the odds of an event happening associated with a one unit change in the value of a predictor.

Pearson goodness-of-fit statistic: A goodness-of-fit test statistic based on comparing the model-based variability in the data to its inherent variability. Under certain conditions it has an asymptotic χ^2 distribution with appropriate degrees of freedom, but is not applicable for this purpose when the number of replications per observation is small.

Prior probabilities: The true (unconditional) probabilities of success and failure. These are required in order to convert estimated probabilities of success from a retrospective study into estimated prospective probabilities.

Probit regression: A regression model for binary data that uses the cumulative distribution function of a normally distributed random variable to generate S-shaped curves, rather than the expit function used in logistic regression.

Prospective design: A study in which sampling of units is not based on the outcome responses, but rather either randomly from the population or perhaps based on predictor(s). Such studies sometimes follow units over long periods of time, observing outcome(s) as they manifest themselves. This is also referred to as a cohort design.

Retrospective design: A study in which sampling of units is based on the response categories that are being studied. This is also referred to as a case-control design.

ROC (Receiver Operating Characteristic) curve: A plot of the fraction of observations classified as successes by a model that are actually successes (the true positive rate) versus the fraction of observations classified as failures that are actually failures (the true negative rate), at various classification cutoff values.

Simpson's paradox: A situation where the marginal direction of association of a variable with a binary response variable is in the opposite direction to the conditional association taking into account membership in groups defined by another variable.

Somers' D: A measure of the ability of a model to separate successes from failures. It is the difference between the proportion of concordant pairs and the proportion of discordant pairs calculated from pairs of estimated success probabilities where one member of the pair is a success and the other is a failure, and a concordant (discordant) pair is one where the estimated probability of success for the actual success is larger (smaller) than that for the actual failure. D ranges from 0 (no separation) to 1 (perfect separation), and is equivalent to the area under the ROC curve through the relationship $D = 2 \times AUR - 1$.

Wald test statistic: A test statistic for the significance of a regression coefficient based on the ratio of the estimated coefficient to its standard error. For large samples the statistic can be treated as a normal deviate.

Multinomial Regression

9.1   Introduction

The formulation of logistic regression in the previous chapter is appropriate for binary response data, but there are situations where it is of interest to model a categorical response that has more than two categories; that is, one that is **polytomous**. For example, in a clinical trial context for a new drug, the responses might be "No side effects," "Headache," "Back pain," and "Dizziness," and the purpose of the study is to see what factors are related to the chances of an individual experiencing one of the possible side effects. Another common situation is the modeling of so-called **Likert-type scale** variable, where (for example) a respondent is asked a question that has response categories "Strongly disagree" – "Disagree" – "Neutral" – "Agree" – "Strongly agree," and the purpose of the study is to see if it is possible to model a person's response as a function of age, gender, political beliefs, and so on.

A key distinction in regression modeling of variables of this type compared to the models discussed in the previous chapters is that in this case the response is explicitly multivariate. If the response variable has K categories, the goal is to model the probability vector $\boldsymbol{\pi} = \{\pi_1, \ldots, \pi_K\}$ as a function of the predictors, so there are $K-1$ numbers being modeled for each observation (the probabilities must sum to 1, so the Kth probability is determined by the others). In this chapter we examine models that generalize logistic regression (which has $K = 2$) to the multiple-category situation. This generalization is accomplished in two distinct ways, depending on whether or not there is a natural ordering to the categories.

9.2   Concepts and Background Material

Consider the ith observed response value $\mathbf{y}_i$ of a K-level categorical variable. The value y_{ik} is the number of replications of the ith observation that fall in the kth category in n_i replications. Just as the binomial random variable was the key underlying distribution for binary logistic regression, in this situation the **multinomial** distribution is key. This is represented $\mathbf{y}_i \sim \text{Mult}(n_i, \boldsymbol{\pi}_i)$. The $n_i = 1$ case is naturally thought of as the situation where the observation has a single response corresponding to one of the categories (for example, "Strongly agree"), while $n_i > 1$ refers to the situation where there are replications for a given set of predictor values and the observation is a vector of length K of counts for each category. Note, however, that in fact $\mathbf{y}_i$ is a vector in both situations, with one entry equal to 1 and all of the others equal to 0 when $n_i = 1$.

As was noted in Section 8.2 for binary response data, there are two aspects of a regression model for a categorical response that need to be addressed: the underlying probability distribution for the response, and the model for the relationship between the probability of each response category and the predictors. The multinomial distribution provides the distribution, but the usefulness of models for the relationship with predictors depends on whether or not there is a natural ordering to the response categories, as is described in the next two sections.

9.2.1 NOMINAL RESPONSE VARIABLE

Consider first the situation where the response variable is **nominal**; that is, there is no natural ordering to the categories. The logistic regression model in this case is the natural generalization from the two-category situation to one based on $K - 1$ different logistic relationships. Let the last (Kth) category represent a baseline category; the model then states that the probability of falling into group k given the set of predictor values $\mathbf{x}$ satisfies

$$\log\left(\frac{\pi_k(\mathbf{x})}{\pi_K(\mathbf{x})}\right) = \beta_{0k} + \beta_{1k}x_1 + \cdots + \beta_{pk}x_p, \qquad k = 1, \ldots, K - 1. \quad (9.1)$$

The model is based on $K-1$ separate equations, each having a distinct set of parameters β_k. Obviously, for the baseline category K, $\beta_{0K} = \beta_{1K} = \cdots = \beta_{pK} = 0$.

The logit form of (9.1) implies that (as before) exponentiating the slope coefficients gives odds ratios, now relative to the baseline category. That is, $e^{\beta_{jk}}$ is the multiplicative change in the odds of being in group k versus being in group K associated with a one unit increase in x_j holding all else in the model fixed. In a situation where there is a naturally defined reference group, it should be chosen as the baseline category, since then the slopes are directly interpretable with respect to that reference. So, for example, in the clinical trial context mentioned earlier, the reference group (and therefore baseline category) would be "No side effects," and the slopes would be interpreted in terms of the odds of having a particular side effect versus having no side effects.

Despite this interpretation of odds ratios relative to the baseline category, the choice of baseline is in fact completely arbitrary from the point of view of the implications of the model. Say category M is instead taken to be the baseline category. From (9.1),

$$
\begin{aligned}
\log\left[\frac{\pi_k(\mathbf{x})}{\pi_M(\mathbf{x})}\right] &= \log\left[\frac{\pi_k(\mathbf{x})/\pi_K(\mathbf{x})}{\pi_M(\mathbf{x})/\pi_K(\mathbf{x})}\right] \\
&= \log\left[\frac{\pi_k(\mathbf{x})}{\pi_K(\mathbf{x})}\right] - \log\left[\frac{\pi_M(\mathbf{x})}{\pi_K(\mathbf{x})}\right] \\
&= (\beta_{0k} + \beta_{1k}x_1 + \cdots + \beta_{pk}x_p) \\
&\quad - (\beta_{0M} + \beta_{1M}x_1 + \cdots + \beta_{pM}x_p) \\
&= (\beta_{0k} - \beta_{0M}) + (\beta_{1k} - \beta_{1M})x_1 + \cdots + (\beta_{p_k} - \beta_{pM})x_p.
\end{aligned}
$$
$$(9.2)$$

That is, the logit coefficients for level k relative to a baseline category M (both intercept and slopes) are the differences between the coefficients for level k relative to baseline K and the coefficients for level M relative to baseline K. Note that there is nothing in this derivation that requires category M to be a baseline category; equation (9.2) applies for any pair of categories.

Note that (9.2) also implies that no matter which category is chosen as baseline, the probabilities of falling in each level as a function of the predictors will not change. Model (9.1) implies the familiar S-shape for a logistic relationship,

$$
\pi_k(\mathbf{x}) = \frac{\exp(\beta_{0k} + \beta_{1k}x_1 + \cdots + \beta_{pk}x_p)}{\sum_{m=1}^{K}\exp(\beta_{0m} + \beta_{1m}x_1 + \cdots + \beta_{pm}x_p)}.
$$
$$(9.3)$$

Equation (9.2) shows that for this model the (log-)odds of one category of the response versus another does not depend on any of the other categories; that is, other possible outcomes are not relevant. This is known as **independence of irrelevant alternatives** (IIA). This is often reasonable, but in some circumstances might not be. Consider, for example, a discrete choice situation, where an individual must choose between, for example, different travel options. A well-known discrete choice example, the so-called "red bus / blue bus" example, illustrates the problem. Suppose a traveler must choose between three modes of transportation: red bus, car, or train. Further, say that the traveler had individual characteristics (income, age, gender, etc.) such that he or she has no preference between the three choices, implying that the traveler's probability of choosing each is 1/3. Now, say an indistinguishable alternative to the red bus, a blue bus, becomes available. It seems reasonable that the two types of buses would be of equal appeal to the traveler, resulting in probability 1/6 of being chosen for each, while the probabilities of the other two possible choices would not change. This, however, is a violation of IIA, since adding the supposedly irrelevant alternative of a blue bus has changed the odds of a traveler choosing the red bus versus a car or a train. For this reason, the nominal logistic regression model should only be used in situations where this sort of effect is unlikely to occur, such as when the different response categories are distinct and dissimilar.

9.2.2 ORDINAL RESPONSE VARIABLE

In many situations, such as the Likert-scaled variable described earlier, there is a natural ordering to the groups, and a reasonable model should take that into account. An obvious approach is to just use ordinary least squares regression, with the group membership numerical identifier as the target variable. This will not necessarily lead to very misleading impressions about the relationships between predictors and the response variable, but there are several obvious problems with its use in general:

1. An integral target variable clearly is inconsistent with continuous (and hence Gaussian) errors, violating one of the assumptions when constructing hypothesis tests and confidence intervals.

2. Predictions from an ordinary regression model are of course nonintegral, resulting in difficulties in interpretation: what does a prediction to category 2.763 mean, exactly?

3. The least squares regression model doesn't address the actual underlying structure of the data, as it ignores the underlying probabilities π completely. A reasonable regression model should provide direct estimates of $\pi(\mathbf{x})$.

4. The regression model implicitly assumes that the numerical codings (for example, 1–2–3–4–5) reflect the actual "distances" from each other, in the sense that group 1 is as far from group 2 as group 2 is from group 3. It could easily be argued that this is not sensible for a given data set. For example, it is reasonable to consider the possibility that a person who strongly agrees with a statement (i.e., group 5) has a somewhat extreme position, and is therefore "farther away" from a person who agrees with

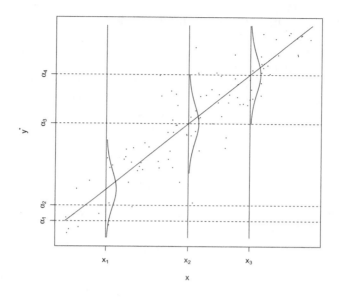

FIGURE 9.1 **Graphical representation of the latent variable model for regression with an ordinal response variable.**

the statement (group 4) than a person who agrees with the statement is from a person who is neutral (group 3). A reasonable regression model should be flexible enough to allow for this possibility.

A way of constructing such a model is through the use of a **latent variable**. The idea is represented in Figure 9.1 in the situation with one predictor, but it generalizes to multiple predictors in the obvious way. Consider the situation where the observed response is a Likert-scaled variable with K categories. The latent variable y^* is an underlying continuous response variable representing the "true" agreement of the respondent with the statement. This variable is not observed, because the respondent is restricted to choosing one of K response categories. To account for this we assume that there is a grid $\{\alpha_0, \ldots, \alpha_K\}$, with $-\infty = \alpha_0 < \alpha_1 < \cdots < \alpha_K = \infty$, such that the observed response y satisfies

$$y = k \text{ if } \alpha_{k-1} < y^* \le \alpha_k.$$

That is, we observe a response in category k when the underlying y^* falls in the kth interval of values.

Note that Figure 9.1 also highlights the linear association between the latent variable y^* and the predictor x. The plotted density curves represent the probability distributions of y^* for specified values of x (x_1, x_2, and x_3, respectively). These curves are centered at different values of y^* because of that linear association. It is this association that ultimately drives the relationship between the probabilities of falling into each of the response categories of y (the observed categorical response) and the predictor, as $\pi_k(x)$ is simply the area under the density curve between α_{k-1} and α_k for given x.

It turns out that if the assumed density of y^* given x follows a logistic distribution, this representation implies a simple model that can be expressed in terms of logits. (The logistic density is similar to the Gaussian, being symmetric but slightly longer-tailed; its cumulative distribution function generates the expit curve.) Define the **cumulative logit** as

$$\mathcal{L}_k(x) = \text{logit}[F_k(x)] = \log\left[\frac{F_k(x)}{1 - F_k(x)}\right], \qquad k = 1, \ldots, K-1,$$

where $F_k(x) = P(y \leq k|x) = \sum_{m=1}^{k} \pi_m(x)$ is the cumulative probability for response category k given x. That is, $\mathcal{L}_k(x)$ is the log odds given x of observing a response less than or equal to k versus greater than k. The **proportional odds model** then says that

$$\mathcal{L}_k(x) = \alpha_k + \beta x, \qquad k = 1, \ldots, K-1.$$

Note that this means that a positive β is associated with increasing odds of being less than a given value k, so a positive coefficient implies increasing probability of being in lower-numbered categories with increasing x; for this reason some statistical packages reverse the signs of the slopes so that a positive slope is consistent with higher values of the predictor being associated with higher categories of the response. This generalizes in the obvious way to multiple predictors,

$$\mathcal{L}_k(\mathbf{x}) = \alpha_k + \beta_1 x_1 + \cdots + \beta_p x_p, \qquad k = 1, \ldots, K-1. \qquad (9.4)$$

The model is called the proportional odds model because it implies that holding all else fixed a one-unit increase in x_j is associated with multiplying the odds of being less than or equal to category k versus being greater than that category by e^{β_j} for any category k.

9.3   Methodology

9.3.1 ESTIMATION

Estimation of the underlying parameters for either of these classes of models can be estimated using maximum likelihood. The log-likelihood is

$$L = \sum_{k=1}^{K} \sum_{y_i = k} \log \pi_{k(i)},$$

where the second summation is over all observations i with response level k, and $\pi_{k(i)}$ is the probability (9.3) for the nominal logistic regression model or the probability implicitly implied by (9.4) for the proportional odds model, respectively, substituting in the predictor values for the ith observation. The resultant estimates of β are efficient for large samples. Note that although the nominal logistic regression has the structure of $K - 1$ separate binary logistic regressions, the underlying relationships are fitted simultaneously.

9.3.2 INFERENCE, MODEL COMPARISONS, AND STRENGTH OF FIT

Many of the inferential and descriptive tools described in Section 8.3.2 for binary logistic regression also apply in the multiple-category case. Output from the fitting of a nominal logistic regression looks very much like a set of binary logistic regression outputs, since the model is defined as a series of logistic regressions with the same baseline category. That is, if for example the baseline category is category K, the output will provide summaries of models for logistic regressions of category 1 versus category K, category 2 versus category K, and so on, resulting in $(K - 1)(p + 1)$ regression parameters. For the logistic regression comparing the kth category to the baseline category Wald tests can be constructed to test the significance of an individual regression coefficient, testing the hypotheses

$$H_0 : \beta_{jk} = 0$$

versus

$$H_a : \beta_{jk} \neq 0.$$

It is also possible to construct a test of the significance of a predictor x_j in all of the regressions, testing

$$H_0 : \beta_{j1} = \cdots = \beta_{jK} = 0$$

versus the alternative that at least one of these coefficients is not zero. The likelihood ratio statistic for testing these hypotheses is the difference between $-2L$ for the model that includes x_j and $-2L$ for the model that does not, and can be compared to a χ^2 critical value on $K - 1$ degrees of freedom. A test for whether all of the slope terms equal zero is a test of the overall significance of the regression, and can be constructed as a likelihood ratio test comparing the fitted model to one without any predictors. If one model is a subset of another model a test of the adequacy of the simpler model is constructed based on the difference between the two $-2L$ values, referenced to a χ^2 distribution with degrees of freedom equal to the difference in the number of estimated parameters in the two models. Different models also can be compared using AIC or AIC_c.

The proportional odds model has a more straightforward representation, being based on a set of $K - 1$ intercepts and a single set of p slope parameters. Wald test statistics can be constructed for the significance of each of these parameters, and a test for whether all of the slope terms equal zero is a test of the overall significance of the regression. Clearly the nominal logistic regression model can be applied to ordinal response data as well, and can provide a check of the validity of the ordinal model, since it is a more general model. For $K > 2$ the proportional odds model is simpler (has fewer parameters) than the nominal logistic regression model, so if it is a reasonable representation of the underlying relationship it would be preferred. The two models can be compared using AIC or AIC_c, and also using a likelihood ratio test. A generalization of (9.4) from constant slopes (β_j) to different slopes for different categories (β_{jk} for the kth category) corresponds to an interaction between the variable x_j and the category; comparing the fits of the two models using a likelihood ratio test is a test of the proportional odds assumption.

Measures of association such as Somers' D (Section 8.3.4) can be constructed for ordinal response models, since in that case the concepts of concordance and discordance are meaningful. A pair of observations from two different categories are concordant if the observation with the lower ordered response value has a lower estimated mean score than the observation with the higher ordered response value, and D is the difference between the concordant and discordant proportions.

If it was desired to classify observations, they would be assigned to the category with largest estimated probability. A classification table is then formed in the same way as when there are two groups (Section 8.3.4), and the two benchmarks $C_{\max}$ and C_{pro} are formed in an analogous way.

9.3.3 LACK OF FIT AND VIOLATIONS OF ASSUMPTIONS

The Pearson (8.6) and deviance (8.7) goodness-of-fit test statistics also generalize to multiple-category regression models. The Pearson statistic is now

$$X^2 = \sum_{i=1}^{n} \sum_{k=1}^{K} \frac{(y_{ik} - n_i \hat{\pi}_{ik})^2}{n_i \hat{\pi}_{ik}},$$

while the deviance is

$$G^2 = 2 \sum_{i=1}^{n} \sum_{k=1}^{K} \left[y_{ik} \log \left(\frac{y_{ik}}{n_i \hat{\pi}_{ik}} \right) \right].$$

Each of these statistics can be compared to a χ^2 distribution on $n(K-1)-p-1$ degrees of freedom, but only if the n_i values are reasonably large (note that as in Chapter 8, n is the number of replicated observations, not the total number of replications). When the number of replications is small, or there are no replications at all ($n_i = 1$ for all i), the Pearson and deviance statistics are not appropriate for testing goodness-of-fit, but Pigeon and Heyse (1999) proposed a test that generalizes the Hosmer-Lemeshow test.

The multivariate nature of the response variable y results in some difficulties in constructing diagnostics for the identification of unusual observations, since there are K different residuals $y_{ik} - n_i \hat{\pi}_{ik}$ for each observation. In the situation with replications it is possible to see whether for any observation any categories have unusually greater or fewer replications than would be expected according to the fitted model. When there is only a single replication, it can be noted that an observation has low estimated probability of falling in its actual category, but for nominal data there is no notion of how "far away" its actual category is from its predicted one. Unusualness for ordinal data is easier to understand, as (for example) an individual with response "Strongly agree" is clearly more unusual if they have a high estimated probability of saying "Strongly disagree" than if they have a high estimated probability of saying "Agree."

9.4 Example — City Bond Ratings

Abzug et al. (2000) gathered data relating the bond rating of a city's general obligation bonds to various factors as part of a study of what types of organizations dominate employment in cities. We focus here on four potential predictors of bond rating: logged population (using natural logs), average total household income, the number of nonprofits among the top ten employers in the city, and the number of for profits among the top ten employers in the city, for 56 large cities that issued general obligation bonds. The bond rating for each city falls into one of the classes AAA, AA, A, or BBB. These classes are ordered from highest to lowest in terms of credit-worthiness.

Figure 9.2 gives side-by-side boxplots of each predictor separated by rating class. There isn't any obvious relationship between logged population and rating or the number of for profit institutions among the top ten employers in the city and rating, while lower average household income and a lower number of nonprofits among the top ten employers are associated with higher credit-worthiness.

The output below summarizes the result of a nominal logistic regression fit based on all four predictors, with the lowest (BBB) rating class taken as the baseline category.

	Estimate	Std. Err	z	Pr(>|z|)
Logit 1: AAA versus BBB				
(Intercept)	12.7268	15.025	0.85	0.397
Logged.population	−0.13988	1.3849	−0.10	0.920
Household.income	−0.00012	0.000166	−0.74	0.456
Nonprofits.in.top.10	−1.58691	0.74459	−2.13	0.033
For.profits.in.top.10	−0.15720	0.38536	−0.41	0.683
Logit 2: AA versus BBB				
(Intercept)	27.6381	11.112	2.49	0.013
Logged.population	−1.6257	0.91953	−1.77	0.077
Household.income	0.00005	0.000107	0.47	0.637
Nonprofits.in.top.10	−1.5765	0.54870	−2.87	0.004

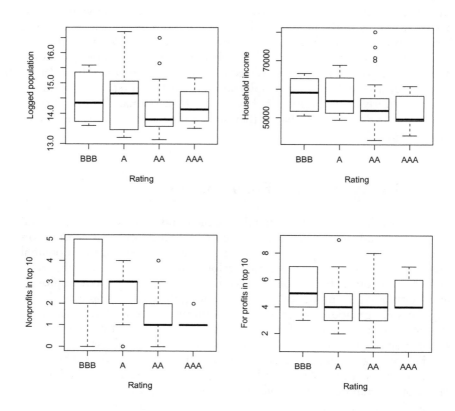

FIGURE 9.2 Side-by-side boxplots for the bond rating data.

```
For.profits.in.top.10          -0.4239    0.31629   -1.34   0.180

Logit 3: A versus BBB
(Intercept)                     7.5600    10.511     0.72    0.470
Logged.population              -0.2661     0.88151   -0.30   0.763
Household.income            -0.000003     0.000110  -0.02   0.981
Nonprofits.in.top.10           -0.4917     0.52392   -0.94   0.348
For.profits.in.top.10          -0.3239     0.31626   -1.02   0.306
```

Most of the Wald statistics are not statistically significant, with only the number of nonprofits among the top ten employers in the city showing up as statistically significant at a .05 level in the comparisons of the cities with AAA ratings versus BBB ratings and those with AA ratings compared to those with BBB ratings. The interpretation of the coefficients is the same as in binary logistic regressions; for example, the slope coefficient for the number of nonprofits in the AAA versus BBB model implies that holding all else fixed an increase of one nonprofit among the top ten employers in a city is associated with multiplying the odds of the city having an AAA rating versus having a BBB rating by $e^{-1.587} = 0.20$, or an estimated 80% decrease. All of the estimated coefficients for this predictor are negative, which is consistent with the marginal relationship that more nonprofits is associated with less chance of having a higher credit rating.

This model seems to be overspecified and should be simplified. Since there are four predictors, there are $2^4 - 1 = 15$ different possible regression models that include at least one predictor. Comparison of all of these models via AIC_c (not shown) implies that only two models should be considered: the model using the number of nonprofits among the top ten employers alone, or the model that also includes logged population. While the simpler model has smaller AIC_c, the model that adds logged population has smaller AIC, and the likelihood ratio test comparing the two models weakly rejects the simpler model ($LR = 7.35$, $df = 3$, $p = .06$). The output below summarizes the fit based on the two-predictor model.

	Estimate	Std. Err	z	Pr(>|z|)
Logit 1: AAA versus BBB				
(Intercept)	17.8415	12.4327	1.44	0.151
Logged.population	−1.00619	0.818426	−1.23	0.219
Nonprofits.in.top.10	−1.70719	0.697162	−2.45	0.014
Logit 2: AA versus BBB				
(Intercept)	23.2681	9.74556	2.39	0.017
Logged.population	−1.28134	0.624626	−2.05	0.040
Nonprofits.in.top.10	−1.44526	0.517454	−2.79	0.005
Logit 3: A versus BBB				
(Intercept)	5.85505	9.42789	0.62	0.535
Logged.population	−0.269707	0.595171	−0.45	0.650
Nonprofits.in.top.10	−0.454743	0.496228	−0.92	0.359

The slopes for the number of nonprofits variable do not change greatly from the model using all four predictors. The only Wald statistic that is statistically significant for the logged population variable is for the comparison of AA versus BBB rating classes, with a larger population associated with less credit-worthiness. Note that since the predictor is in the natural log scale, the slope is an elasticity for the odds; that is, holding the number of nonprofits fixed, a 1% increase in population is associated with an estimated 1.28% decrease in the odds of a city being rated AA versus BBB. The overall statistical significance of the model is high, with $LR = 20.6$ ($df = 6$, $p = .002$).

This analysis, of course, does not take into account the natural ordering of the rating classes. Here is output for a proportional odds fit to the data based on all four of the predictors:

```
                        Coef      S.E.      Z    Pr(>|Z|)
alpha(1)               10.9797   5.2474   2.09   0.0364
alpha(2)               14.8816   5.4321   2.74   0.0062
alpha(3)               16.7583   5.5639   3.01   0.0026
Logged.population      -0.8075   0.4459  -1.81   0.0701
Household.income        0.0000   0.0000  -0.02   0.9815
Nonprofits.in.top.10   -1.0773   0.2798  -3.85   0.0001
For.profits.in.top.10  -0.0845   0.1557  -0.54   0.5874
```

Note that the software fitting this model reverses the signs, so (for example) the negative coefficient for the number of nonprofits among the top ten employers implies that more nonprofits are associated with a lower credit rating (holding all else fixed). The only two variables close to statistical significance are again logged population and the number of nonprofits among the top ten employers. Comparison of all of the possible models via AIC_c implies that the only model supported is that based on logged population and the number of nonprofits among the top ten employers:

```
                        Coef      S.E.      Z    Pr(>|Z|)
alpha(1)               10.7301   4.8965   2.19   0.0284
alpha(2)               14.6202   5.0974   2.87   0.0041
alpha(3)               16.4908   5.2338   3.15   0.0016
Logged.population      -0.8205   0.3384  -2.42   0.0153
Nonprofits.in.top.10   -1.0670   0.2744  -3.89   0.0001
```

The overall regression is highly statistically significant ($LR = 19.7$, $df = 2$, $p < .0001$). The implications of the model are similar to those of the nominal model (higher population and more nonprofits are associated with lower credit rating holding all else fixed), but this model is much more parsimonious, being based on only 5 parameters rather than 9. This is reflected in the AIC_c values (AIC_c for the ordinal model is 10.5 lower than that of the nominal model), and also in the likelihood ratio test comparing the two models, which does not come close to rejecting the simpler ordinal model ($LR = 0.9$, $df = 4$, $p = 0.93$).

Somers' $D = 0.52$ for this model, suggesting moderate separation between the rating classes. This is also reflected in the classification table (the corresponding table for the nominal logistic regression model based on these two variables is unsurprisingly very similar).

		BBB	A	AA	AAA	
	BBB	2	2	2	0	6
Actual	A	0	6	6	0	12
result	AA	0	2	31	0	33
	AAA	0	0	5	0	5
		2	10	44	0	

Predicted result

The model correctly classifies 69.6% of the cities, which is only moderately greater than $C_{max} = 58.9\%$ and $C_{pro} = 63.1\%$. Two of the cities are misclassified to two rating classes higher than they were (Buffalo and St. Louis both had ratings of BBB but were classified to an AA rating, because of a low number of nonprofits among the top ten employers [Buffalo] and a large population [St. Louis], respectively). Even more troubling, none of the cities are classified to the AAA group.

Omitting Buffalo and St. Louis from the data does not change the fitted models very much. The proportional odds model based on logged population and number of nonprofits among the top ten employers does now classify one city to the AAA rating class, but it is an incorrect classification, so that is not a point in its favor. An alternative approach could be based on Simonoff (1998b), in which a smoothed version of the cross-classification of bond rating and number of nonprofits is used to argue that the relationship between rating and nonprofits is not monotone (both less and more nonprofits than one among the top ten employers being associated with lower bond rating). This suggests accounting for this pattern in the regression by fitting a quadratic relationship with number of nonprofits, and if that is done the squared predictor is in fact statistically significant. Classification accuracy does not improve based on this model, however.

9.5 Summary

In this chapter we have generalized the application of logistic regression to response data with more than two categories. Nominal logistic regression is based on the principle of choosing a baseline category and then simultaneously estimating separate (binary) logistic regressions for each of the other categories versus that baseline. This is a flexible approach, but can involve a large number of parameters if K is large.

The nominal logistic regression model requires the assumption of independence of irrelevant alternatives, an assumption that can easily be violated in discrete choice models. There is a large literature on tests for IIA and extensions and generalizations of multiple category regression models that are appropriate in the discrete choice framework. See Train (2009) for more details.

If the response variable has naturally ordered categories, it is appropriate to explore models that take that into account, as they can often provide parsimonious representations of the relationships in the data. These models are generally underutilized in practice, as analysts tend to just use ordinary (least squares) linear regression with the category number as the response value to analyze these data. This is not necessarily a poor performer if the number of categories is very large, but can be very poor for response variables with a small number of categories. The proportional odds model is a standard first approach, but it is not the only possibility. Other possible models include ones based on cumulative probits, adjacent-categories logits, or continuation ratios rather than cumulative logits. Chapter 10 of Simonoff (2003) provides more extensive discussion of these models.

KEY TERMS

Cumulative logit: The logit based on the cumulative distribution function, representing the log-odds of being at a specific level or lower versus being at a higher level.

Independence of irrelevant alternatives (IIA): The property that the odds of being in one category of a multiple-category response variable versus another category of the variable depends only on the two categories, and not on any other categories.

Latent variable: An underlying continuous response variable representing the "true" agreement of a respondent with a statement, which is typically unobservable. More generally, any such variable that is only observed indirectly through a categorical variable that reports counts falling into (unknown) intervals of the variable.

Likert-type scale variable: A variable typically used in surveys in which a respondent reports his or her level of agreement or disagreement on an ordered scale. The scale is typically thought of as reflecting an underlying latent variable.

Multinomial random variable: A discrete random variable that takes on more than two prespecified values. This is a generalization of the binomial variable, which takes on only two values.

Nominal variable: A categorical variable where there is no natural ordering of the categories.

Ordinal variable: A categorical variable where there is a natural ordering of the categories.

Polytomous variable: A categorical response that has more than two categories.

Proportional odds model: A regression model based on cumulative logits of an ordinal response variable that hypothesizes an equal multiplicative effect of a predictor on the odds of being less than or equal to category k versus being greater than that category for any category k, holding all else fixed. It is consistent with an underlying linear relationship between a latent variable and the predictors if the associated error term follows a logistic distribution.

CHAPTER TEN

Count Regression

10.1 Introduction

The previous two chapters focused on situations where least squares estimation is not appropriate because of the special nature of the response variable, with Chapter 8 exploring binary logistic regression models for binomial response data and Chapter 9 outlining multiple-category logistic regression models for multinomial response data. Another situation of this type is the focus of this chapter: regression models when the response is a count, such as (for example) attempting to model the expected number of homicides in a small town for a given year or the expected number of lifetime sex partners reported by a respondent in a social survey.

Just as was true for binary data, the use of least squares is inappropriate for data of this type. A linear model implies the possibility of negative estimated mean responses, but counts must be nonnegative. Counts can only take on (nonnegative) integer values, which makes them inconsistent with Gaussian errors. Further, it is often the case that count data exhibit heteroscedasticity, with larger variance accompanying larger mean.

In this chapter we examine regression models for count data. The workhorse random variable in this context is the Poisson random variable, and we first describe its properties. We then highlight the many parallels between count regression models and binary regression models by showing how both (and also Gaussian-based linear regression) are special cases of a broad class of models called generalized linear models. This general formulation provides a framework for inference in many regression situations, including count regression models.

Although the Poisson random variable provides the basic random structure for count regression modeling, it is not flexible enough to handle all count regression problems. For this reason we also discuss various generalizations of Poisson regression to account for more variability than expected (overdispersion), and greater or fewer observed specific numbers of counts (often greater or fewer zero counts than expected), which can be useful for some data sets.

10.2   Concepts and Background Material

10.2.1 THE POISSON RANDOM VARIABLE

The normal distribution is not an appropriate choice for count data for the reasons noted earlier. The standard distribution for a count is a Poisson distribution. The reason for this is that the Poisson is implied by a general model for the occurrence of random events in time or space. In particular, if ν is the rate at which events occur per unit time, under reasonable assumptions for the occurrence of events in a small time interval, the total number of independent events that occur in a period of time of length t has a Poisson distribution with mean $\mu = \nu t$. The Poisson distribution, represented as $y \sim \text{Pois}(\mu)$, is completely determined by its mean μ, as its variance also equals μ.

The Poisson random variable is closed under summation, in the sense that a sum of independent Poissons is itself Poisson with mean equal to the sum of the underlying means. Since this means that a Poisson with mean μ is the sum of μ independent Poissons of mean 1, the Central Limit Theorem implies that as μ gets larger a Poisson random variable is approximately normal.

The Poisson distribution also has important connections with two other discrete data distributions. If the number of successes in n trials is binomially distributed, with the number of trials $n \to \infty$ and the probability of success $p \to 0$ such that $np \to \mu$, the distribution of the number of successes is approximately Poisson with μ. This means that the Poisson is a good choice to model the number of rare events; that is, ones that are unlikely to occur in any one situation (since p is small), but might occur in a situation with many independent trials (that is, n is large).

The Poisson random variable also has a close connection to the multinomial. If K independent Poisson random variables $\{n_1, \ldots, n_K\}$ are observed, each with mean μ_i, their joint distribution conditional on the total number of counts $\sum_j n_j$ is multinomial with probability $\pi_i = \mu_i/(\sum_j \mu_j)$. This connection turns out to be particularly important in the analysis of tables of counts (contingency tables).

10.2.2 GENERALIZED LINEAR MODELS

As was noted in Section 8.2, a regression model must specify several things: the distribution of the value of the response variable y_i (the so-called *random component*), the way that the predicting variables combine to relate to the level of y_i (the *systematic component*), and the connection between the random and systematic components (the *link function*). The **generalized linear model** is a family of models that provides such a framework for a very wide set of regression problems. Specifically, the random component requires that the distribution of y_i comes from the exponential family and the systematic component specifies that the predictor variables relate to the level of y as a linear combination of the predictor values (a *linear predictor*). The link function then relates this linear predictor to the mean of y.

The Gaussian linear model (1.1) is an example of a generalized linear model, with identity link function and a normal distribution as the random component. Logistic regression is also an example of a generalized linear model, with the binomial distribution being the random component and the logit function (8.1) defining the link function.

The Poisson regression model is also a member of the generalized linear model family. Since the mean of a Poisson random variable must be nonnegative, a natural link function in this case (and the one that is standard) is the log link,

$$\log \mu_i \equiv \log[E(y_i)] = \beta_0 + \beta_1 x_{1i} + \cdots + \beta_p x_{pi}.$$

Thus, the Poisson regression model is an example of a log-linear model. By convention natural logs are used rather than common logs. Note that this is a semilog model for the mean of y,

$$\mu_i = e^{\beta_0 + \beta_1 x_{1i} + \cdots + \beta_p x_{pi}}, \tag{10.1}$$

so the slope coefficients have the usual interpretation as semielasticities, with a one unit change in x_j associated with multiplying the expected response by e^{β_j}, holding all else in the model fixed.

10.3   Methodology

10.3.1 ESTIMATION AND INFERENCE

Maximum likelihood is typically used to estimate the parameters of generalized linear models, including the Poisson regression model. The log-likelihood takes the form

$$L = \sum_{i=1}^{n} (y_i \log \mu_i - \mu_i),$$

where μ_i satisfies (10.1). Just as was true for logistic regression, the maximizer of this function (the maximum likelihood estimate) takes the approximate form of a weighted least squares (IRWLS) estimate with weight and corresponding error variance for each observation that depends on the parameter estimates.

Inference proceeds in a completely analogous way to the situation for logistic regression. Specifically:

- A test comparing any two models where one is a simpler special case of the (more general) other is the likelihood ratio test LR, where

$$LR = 2(L_{\text{general}} - L_{\text{simpler}}).$$

This is compared to a χ^2_d critical value, where d is the difference in the number of parameters fit under the two models. This applies also to a test of the overall significance of the regression, where the general model is the model using all of the included predictors, and the simpler model uses only the intercept.

- Assessment of the statistical significance of any individual regression coefficient β_j also can be based on a LR test, but is more typically based on the Wald test

$$z_j = \frac{\hat{\beta}_j}{\widehat{s.e.}(\hat{\beta}_j)}.$$

Similarly, an asymptotic $100 \times (1 - \alpha)\%$ confidence interval for β_j has the form $\hat{\beta}_j \pm z_{\alpha/2}\widehat{s.e.}(\hat{\beta}_j)$.

- Models can be compared using AIC, with

$$AIC = -2L + 2(p + 1).$$

As before, the corrected version has the form

$$AIC_c = AIC + \frac{2(p + 2)(p + 3)}{n - p - 3}.$$

Note that (as was true for logistic regression models) the theory under-lying AIC_c does not apply to Poisson regression, but the criterion has proven to be useful in practice.

• If the values of $E(y_i)$ are large, the Pearson (X^2, the sum of squared Pearson residuals) and deviance (G^2) goodness-of-fit statistics can be used to assess goodness-of fit. Each is compared to a χ^2 distribution on $n-p-1$ degrees of freedom. In the Poisson regression situation

$$X^2 = \sum_{i=1}^{n} \frac{(y_i - \hat{\mu}_i)^2}{\hat{\mu}_i}$$

and

$$G^2 = 2 \sum_{i=1}^{n} \left[y_i \log \left(\frac{y_i}{\hat{\mu}_i} \right) - (y_i - \hat{\mu}_i) \right].$$

• Diagnostics are approximate, based on the IRWLS representation of the maximum likelihood regression estimates.

10.3.2 OFFSETS

The Poisson regression model is based on modeling the expected number of occurrences of some event as a function of different predictors, but sometimes this is not the best choice to model. In some situations what is of interest is the expected *rate* of occurrence of an event, rather than the expected *number* of occurrences. Here the rate is appropriately standardized so as to make the values comparable across observations. So, for example, if one was analyzing marriages by state, the actual number of marriages would not be of great interest, since its strongest driver is simply the population of the state. Rather, it is marriage rate (marriages per 100,000 population, for example) that is comparable across states, and should therefore be the focus of the analysis. This means that the appropriate model is

$$y_i \sim \text{Pois}[k_i \times \exp(\beta_0 + \beta_1 x_{1i} + \cdots + \beta_p x_{pi})],$$

where k_i is the standardizing value (such as population) for the ith observa-tion, and

$$\exp(\beta_0 + \beta_1 x_{1i} + \cdots + \beta_p x_{pi})$$

now represents the mean rate of occurrence, rather than the mean number of occurrences. Equivalently, the model is

$$y_i \sim \text{Pois}\{\exp[\beta_0 + \beta_1 x_{1i} + \cdots + \beta_p x_{pi} + \log(k_i)]\}.$$

This means that a Poisson rate model can be fit by including $\log(k_i)$ (the so-called **offset**) as a predictor in the model, and forcing its coefficient to equal one.

10.4   Overdispersion and Negative Binomial Regression

As was noted earlier, the Poisson random variable is restrictive in that its variance equals its mean. Often in practice the observed variance of count data is larger than the mean; this is termed **overdispersion**. The most common cause of this is unmodeled heterogeneity, where differences in means between observations are not accounted for in the model. Note that this also can occur for binomial data (and hence in logistic regression models), since the binomial random variable also has the property that its variance is exactly determined by its mean. There are specific tests designed to identify overdispersion, but often the standard goodness-of-fit statistics X^2 and G^2 can identify the problem. The presence of overdispersion should not be ignored, since even if the form of the fitted log-linear model is correct, not accounting for overdispersion leads to estimated variances of the estimated coefficients that are too small, making confidence intervals too narrow and p-values of significance tests too small. In particular, the estimated standard errors of the estimated coefficients are too small by the same factor as the ratio of the true standard deviation of the response to the estimated one based on the Poisson regression. So, for example, if the true standard deviation of y is 20% larger than that based on the Poisson regression, the estimated standard errors should also be 20% larger to reflect this.

10.4.1 QUASI-LIKELIHOOD

A simple correction for this effect is through the use of **quasi-likelihood estimation**. Quasi-likelihood is based on the principle of assuming a mean and variance structure for a response variable without specifying a specific distribution. This leads to a set of estimating equations that are similar in form to those for maximum likelihood estimation, and hence a similar estimation strategy based on IRWLS.

Consider a count regression model that posits a log-linear model for the mean (as is the case for Poisson regression), but a variance that is a simple multiplicative inflation over the equality of mean and variance assumed by the Poisson,

$$V(y_i) = \mu_i(1 + \alpha).$$

In this case the quasi-likelihood estimating equations are identical to the Poisson regression maximum likelihood estimating equations, so standard Poisson regression software can be used to estimate β. Since

$$\frac{V(y_i)}{\mu_i} = E\left[\frac{(y_i - \mu_i)^2}{\mu_i}\right] = 1 + \alpha,$$

a simple estimate of $1 + \alpha$ is

$$\widehat{1 + \alpha} = \frac{1}{n - p - 1} \sum_{i=1}^{n} \frac{(y_i - \hat{\mu}_i)^2}{\hat{\mu}_i},$$

which is the Pearson statistic X^2 divided by its degrees of freedom. Thus, quasi-likelihood corrects for overdispersion by dividing the Wald statistics from the standard Poisson regression output by $\sqrt{1 + \alpha} = \sqrt{X^2/(n - p - 1)}$.

Model selection criteria also can be adapted to this situation. The **quasi-AIC** criterion $QAIC$ takes the form

$$QAIC = \frac{-2L}{1 + \hat{\alpha}} + 2(p + 1),$$

while the corresponding bias-corrected version is

$$QAIC_c = QAIC + \frac{2(p + 3)(p + 4)}{n - p - 4}$$

(note that the correction in $QAIC_c$ is slightly different than that for AIC_c, since α is estimated here). Applying these criteria for model selection is not straightforward. All of the $QAIC$ (or $QAIC_c$) values must be calculated using the same value of $\hat{\alpha}$ to be comparable to each other. One strategy is to determine $\hat{\alpha}$ from the most complex model available, and then calculate the model selection criterion for each model using that value. Alternatively, $\hat{\alpha}$ could be chosen based on the "best" Poisson regression model according to AIC (or AIC_c).

10.4.2 NEGATIVE BINOMIAL REGRESSION

An alternative strategy to address overdispersion is to fit a regression model that is based on a random component that (unlike the Poisson distribution) allows for overdispersion. The most common such distribution is the **negative binomial** distribution. The negative binomial arises in several different ways, but one in particular is most relevant here. The standard Poisson regression model assumes that $y_i \sim \text{Pois}(\mu_i)$, with μ_i a fixed mean value that is a function of the predictor values. If instead μ_i is a random variable, this results in unmodeled heterogeneity, and hence overdispersion. The negative binomial random variable arises if μ_i follows a Gamma distribution, and its variance satisfies

$$V(y_i) = \mu_i(1 + \alpha\mu_i).$$

Note that unlike the quasi-likelihood situation discussed in the previous section, here the variance is a function of the square of the mean, not the mean. The Poisson random variable corresponds to the negative binomial with $\alpha = 0$. Some statistical packages parameterize the negative binomial using $\theta = 1/\alpha$, so in that case as $\theta \to \infty$ the negative binomial becomes closer to the Poisson.

The parameters of a negative binomial regression are estimated using maximum likelihood, based on the log-likelihood

$$L = \sum_{i=1}^{n} \log \Gamma(y_i + \theta) - n \log \Gamma(\theta)$$

$$+ \sum_{i=1}^{n} \left[\theta \log \left(\frac{\theta}{\theta + \mu_i} \right) + y_i \log \left(\frac{\mu_i}{\theta + \mu_i} \right) \right],$$

where $\Gamma(\cdot)$ is the gamma function

$$\Gamma(z) = \int_0^{\infty} e^{-t} t^{z-1} dt.$$

The usual IRWLS-based Wald tests and diagnostics apply here. AIC or AIC_c can be used to compare negative binomial fits to each other, as well as to compare negative binomial to Poisson fits, based on the appropriate log-likelihoods and taking into account the additional α parameter.

10.5 Example — Unprovoked Shark Attacks in Florida

The possibility of an unprovoked attack by a shark was the central theme of the movie "Jaws," but just how likely is that to happen? The Florida Program for Shark Research at the Florida Museum of Natural History maintains the International Shark Attack File (ISAF), a list of unprovoked attacks of sharks on humans around the world. The data examined here cover the years 1946 through 2011 for the state of Florida, an area of particular interest since almost all residents and visitors are within 90 minutes of the Atlantic Ocean or Gulf of Mexico. The data for 1946 to 1999 come from Simonoff (2003), supplemented by information for later years given on the International Shark Attack File website. The response variable is the number of confirmed unprovoked shark attacks in the state each year, and potential predictors are the resident population of the state and the year.

Figure 10.1 gives scatter plots of the number of attacks versus year and population. As would be expected, a higher population is associated with a larger number of attacks. Also, there is a clear upwards trend over time. It is also apparent that the amount of variability in attacks increases with the number of attacks, as would be consistent with a Poisson (or negative binomial) random variable.

The Poisson regression output is as follows.

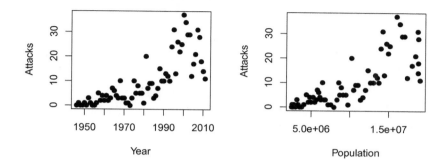

FIGURE 10.1 Scatter plots for the Florida shark attack data.

```
Coefficients:
               Estimate Std. Error  z value Pr(>|z|)
(Intercept) -2.092e+02  5.400e+01   -3.874 0.000107 ***
Year         1.077e-01  2.775e-02    3.880 0.000104 ***
Population  -2.051e-07  9.468e-08   -2.166 0.030327 *
---
Signif. codes:
   0 '***' 0.001 '**' 0.01 '*' 0.05 '.' 0.1 ' ' 1

    Null deviance: 640.2  on 65  degrees of freedom
Residual deviance: 195.8  on 63  degrees of freedom
```

Each of the predictors is statistically significant. The coefficient for Year implies that given population the expected number of attacks is estimated to be increasing 11% per year ($e^{.108} = 1.11$). It should be clear, however, that given the great change in population in Florida since World War II this is a situation where modeling attack rates is more meaningful than modeling number of attacks. Figure 10.2 gives a plot of attack rate versus year, and it is apparent that there is still an increasing trend over time.

Output for a Poisson regression model on year taking logged population as an offset is given below.

```
Coefficients:
            Estimate Std. Error z value Pr(>|z|)
(Intercept) -55.23413   5.30402 -10.414  < 2e-16 ***
Year          0.02080   0.00266   7.819 5.34e-15 ***
---
```

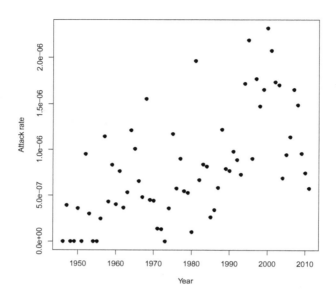

FIGURE 10.2 Plot of attack rate versus year for the Florida shark attack data.

```
Signif. codes:
  0 '***' 0.001 '**' 0.01 '*' 0.05 '.' 0.1 ' ' 1

    Null deviance: 254.89  on 65  degrees of freedom
Residual deviance: 187.55  on 64  degrees of freedom
```

The relationship is strongly statistically significant, and it implies an estimated 2.1% annual increase in shark attack rates. Figure 10.3, however, indicates problems with the model. The time series plot of the standardized Pearson residuals gives evidence of a downward trend in the residuals for the last ten years of the sample (that is, 2002 through 2011), and the Cook's distances are noticeably larger for the last two years.

This pattern is, in fact, not surprising. The ISAF Worldwide Shark Attack Summary (Burgess, 2012) noted that there were notable slow-downs in local economies after the September 11, 2001 terrorist attacks in the United States, which were exacerbated by the 2008-2011 recession and the very active tropical seasons in Florida in 2004, 2005, and 2006. As a result fewer people were entering the water during this time period. Further, extensive media coverage of the "do's and don't's" of shark-human interactions may have led to people reducing their interactions with sharks.

This suggests that the shark attack rate pattern might be different pre- and post-9/11. This can be handled easily within the Poisson regression frame-

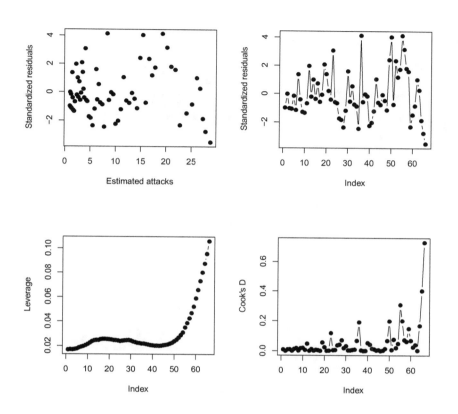

FIGURE 10.3 Diagnostic plots for the Poisson regression model for attack rate using `Year` as the predictor for the Florida shark attack data.

work by adding an indicator variable for the post-9/11 years and the product of that indicator and the year as predictors. The resultant output is as follows.

```
Coefficients:
                 Estimate Std. Error  z value Pr(>|z|)
(Intercept)    -85.486725   7.997612  -10.689  < 2e-16 ***
Year             0.036079   0.004024    8.967  < 2e-16 ***
Post.911       215.593383  49.542971    4.352 1.35e-05 ***
Year:Post.911   -0.107741   0.024703   -4.361 1.29e-05 ***
---
Signif. codes:
  0 '***' 0.001 '**' 0.01 '*' 0.05 '.' 0.1 ' ' 1

    Null deviance: 254.89  on 65  degrees of freedom
Residual deviance: 146.31  on 62  degrees of freedom
```

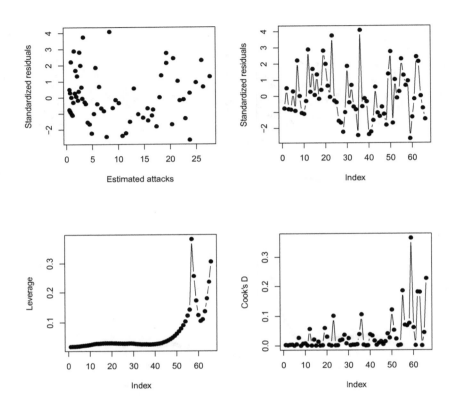

FIGURE 10.4 Diagnostic plots for the Poisson regression model for attack rate based on different relationships with time for the pre- and post-9/11 time periods for the Florida shark attack data.

This model is a clear improvement over the pooled regression line, as all of the predictors are highly statistically significant. A likelihood-ratio test comparing the pooled model to the model based on two separate lines is $LR = 41.24$ on 2 degrees of freedom, which is very highly statistically significant. The model implies an estimated 3.7% annual increase in shark attack rates from 1946-2001 ($e^{.036} = 1.037$), but an estimated 6.9% annual *decrease* in attack rates from 2002-2011 ($e^{.0361 - .1077} = .931$). Diagnostic plots for this model (Figure 10.4) also do not indicate any systematic problems with the model.

Unfortunately, the deviance of $G^2 = 146.3$ on 62 degrees of freedom still indicates lack of fit (the Pearson statistic $X^2 = 145.8$ is similar). Since (based on diagnostic plots) the lack of fit does not seem to come from mis-specification of the expected shark attack rate, this suggests that it is reflecting overdispersion, which would not be surprising given changes in tourism patterns during this 65-year period. One approach to accounting for this is quasi-likelihood, in which the standard errors of the estimated coefficients are multiplied by $\sqrt{145.8/62} = 1.53$. If this is done all of the absolute Wald statistics for the slopes are still greater than 2.8 (corresponding to p-values less than .004), so while statistical significance is weaker, the existence of two different patterns for the pre- and post-9/11 time periods is still supported.

Alternatively, a negative binomial regression model can be fit to the data. Output is given below.

```
Coefficients:
                Estimate  Std. Error  z value  Pr(>|z|)
(Intercept)    -76.600673  10.808080   -7.087  1.37e-12 ***
Year             0.031600   0.005453    5.795  6.82e-09 ***
Post.911       212.958313  94.149355    2.262    0.0237 *
Year:Post.911   -0.106378   0.046933   -2.267    0.0234 *
---
Signif. codes:
  0 '***' 0.001 '**' 0.01 '*' 0.05 '.' 0.1 ' ' 1

    Null deviance: 124.824  on 65  degrees of freedom
Residual deviance:  77.493  on 62  degrees of freedom

         Theta:  7.73
      Std. Err.:  3.00
```

The resultant model has $\hat{\theta} = 7.73$ (or equivalently $\hat{\alpha} = 0.129$), and even after accounting for the overdispersion still supports the existence of different patterns for the two time periods. Diagnostic plots (Figure 10.5) are similar to those for the Poisson regression fit, except that the residuals and Cook's distances are noticeably smaller for the negative binomial fit, indicating that the lack of fit has been addressed to a large extent. Here $G^2 = 77.5$ ($p = .09$) and $X^2 = 71.7$ ($p = .19$), each on 62 degrees of freedom, indicating a moderately reasonable fit.

In all but one year after 1955 there was at least one unprovoked shark attack in Florida, meaning that a direct fitting of a log-linear model for attack rates is possible by taking logs of the attack rate variable and using least squares. This attempt to avoid count regression, while incorrect, results in estimated coefficients not very dissimilar from the Poisson and negative binomial fits, although implications of inference are different. This is not even a consideration, however, in a situation where there are many zero counts among the responses. Consider, for example, modeling the number of annual fatalities from unprovoked shark attacks in Florida. In this case the response only takes on the values 0, 1, or 2, and in more than 80% of the years there were no fatalities. Taking logs is not feasible here, but a log-linear count regression model can be easily fit. Here is output for a Poisson regression model for per capita fatality rate on year and the number of attacks in that

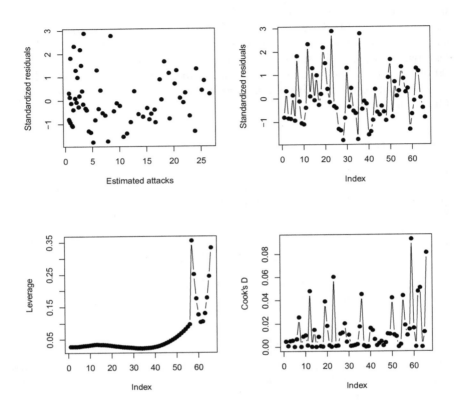

FIGURE 10.5 Diagnostic plots for the negative binomial regression model for attack rate based on different relationships with time for the pre- and post-9/11 time periods for the Florida shark attack data.

year (there is no evidence of different patterns pre- and post-9/11, which is to be expected, since lower tourism rates are already reflected in fewer attacks):

```
Coefficients:
            Estimate Std. Error z value Pr(>|z|)
(Intercept) 95.84120   52.15408   1.838   0.0661 .
Year        -0.05772    0.02657  -2.173   0.0298 *
Attacks      0.07022    0.04164   1.686   0.0917 .
---
Signif. codes:
  0 '***' 0.001 '**' 0.01 '*' 0.05 '.' 0.1 ' ' 1

    Null deviance: 47.586  on 65  degrees of freedom
Residual deviance: 42.419  on 63  degrees of freedom
```

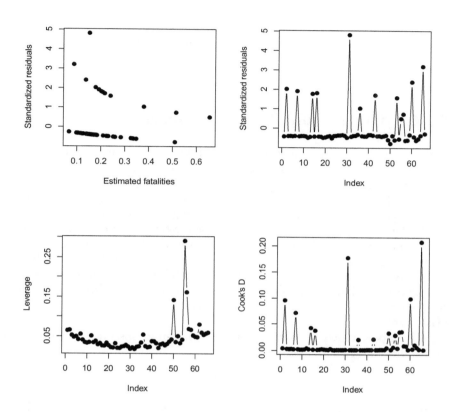

FIGURE 10.6 Diagnostic plots for the Poisson regression model for fatality rate based on year and number of attacks for the Florida shark fatalities data.

The model implies more fatalities in years with more attacks, and given the number of attacks, a decreasing trend in fatality rate of roughly 5.6% annually ($e^{-.0577} = .944$), perhaps reflecting more understanding of how to handle shark attacks among residents and tourists. Diagnostic plots (Figure 10.6) indicate one pronounced outlier (1976, in which there were 2 fatalities in 5 attacks), and omitting that observation strengthens the observed relationships slightly:

```
Coefficients:
              Estimate  Std. Error  z value  Pr(>|z|)
(Intercept)  116.55824    59.28918    1.966    0.0493 *
Year          -0.06843     0.03022   -2.264    0.0235 *
Attacks        0.09389     0.04578    2.051    0.0403 *
---
```

```
Signif. codes:
  0 '***' 0.001 '**' 0.01 '*' 0.05 '.' 0.1 ' ' 1
```

```
    Null deviance: 41.163  on 64  degrees of freedom
Residual deviance: 35.200  on 62  degrees of freedom
```

10.6   Other Count Regression Models

In this section we briefly mention some useful generalizations of these models. It is often the case that more observations with zero counts than expected occur in the observed data. This can be represented through the use of a mixture model, where an observed y_i is modeled as coming from either a point mass at zero (a constant zero) or the underlying count distribution (for example Poisson or negative binomial). These are termed zero-inflated count regression models. In the **zero-inflated Poisson (ZIP)** model, for example, a count is generated from one of two processes: the usual Poisson random variable with mean $\mu_i = \exp(\mathbf{x}_i'\boldsymbol{\beta})$ with probability $1 - \psi_i$, and zero with probability ψ_i. Note that this means that an observation with $y_i = 0$ can come from either the underlying Poisson random variable or from the point mass at zero.

The model allows ψ_i to vary from observation to observation, with the probability modeled to satisfy

$$\psi_i = F(\mathbf{z}_i'\boldsymbol{\gamma}),$$

where $F(\cdot)$ is a cumulative distribution function, $\mathbf{z}_i$ is a set of predictors for the probability of coming from the point mass distribution, and $\boldsymbol{\gamma}$ is a set of parameters corresponding to the predictors. The distribution function F is typically taken to be either the standard normal distribution (a probit model) or the logistic distribution (a logit model). The z variables can include some (or all) of the x's, and can include other variables. The parameters β and γ can be estimated using maximum likelihood. The conditional mean of the target variable is

$$E(y_i|\mathbf{x}_i, \mathbf{z}_i) = \mu_i(1 - \psi_i),$$

while the conditional variance is

$$V(y_i|\mathbf{x}_i, \mathbf{z}_i) = \mu_i(1 - \psi_i)(1 + \mu_i\psi_i),$$

implying overdispersion relative to the Poisson if $\psi_i > 0$. The need for the ZIP regression model over the Poisson regression model can be assessed using an information criterion, such as AIC_c. One simple variation of the model takes the z's to be identical to the x's, and assumes $\boldsymbol{\gamma} = \tau\boldsymbol{\beta}$ for some τ.

If the z and x predictors overlap, interpretation of the estimated parameters $\hat{\boldsymbol{\beta}}$ is not straightforward. For example, if a β coefficient corresponding to a particular predictor is positive, then larger values of that predictor are associated with a larger mean in the Poisson part of the distribution. If, however, this variable has a positive γ coefficient, then larger values are associated with a higher probability of coming from the point mass distribution, lowering the overall expected value of the target variable. These two tendencies can combine in complex ways. If a ZIP formulation with two distinct (nonoverlapping) processes for y is a reasonable representation of reality, things are easier, as then the β coefficients have the usual log-linear interpretation, only now given that the observation comes from the Poisson part of the distribution. These models also can be generalized to incorporate inflation at count values other than zero in the analogous way.

Negative binomial regression also can be modified to allow for zero-count inflation. The **zero-inflated negative binomial (ZINB)** model uses a negative binomial distribution in the mixture rather than a Poisson, but otherwise the formulation is the same. The conditional mean of the response is identical to that for the ZIP model,

$$E(y_i|\mathbf{x}_i, \mathbf{z}_i) = \mu_i(1 - \psi_i),$$

while the conditional variance is

$$V(y_i|\mathbf{x}_i, \mathbf{z}_i) = \mu_i(1 - \psi_i)[1 + \mu_i(\psi_i + \alpha)].$$

If $\psi_i = 0$ this is the standard negative binomial regression model variance, but overdispersion relative to the negative binomial occurs if $\psi_i > 0$. The ZIP model is a special case of the ZINB model (with $\alpha = 0$), so the usual tests can be constructed to test against the one-sided alternative $\alpha > 0$. It should be noted that misspecification of the nondegenerate portion of the model (that is, Poisson or negative binomial) is more serious in the zero-inflated context than in the standard situation, since (unlike in the standard case) parameter estimators assuming a Poisson distribution are not consistent if the actual distribution is negative binomial.

A different situation is one where the observed count response variable is truncated from below, with no values less than a particular value observed. This most typically occurs where truncation occurs at zero. An example of this would be where the response variable is the number of items purchased by a customer among a set of customers on a given day; in this case only customers that purchased at least one item appear in the data, so $y_i = 0$ is not possible. In this case the **zero-truncated** Poisson regression is based on the density

$$f(y_i; \mu_i) = \frac{f_{\text{Pois}}(y_i; \mu_i)}{(1 - e^{-\mu_i})}, \qquad y_i = 1, 2, \ldots,$$

with μ_i following the usual logarithmic link with predictors $x_1, \ldots, x_p$. The mean of the target variable is

$$E(y_i) = \frac{\mu_i}{1 - e^{-\mu_i}}, \tag{10.2}$$

while the variance is

$$V(y_i) = \left[\frac{\mu_i}{1 - e^{-\mu_i}}\right]\left[1 - \frac{\mu_i e^{-\mu_i}}{1 - e^{-\mu_i}}\right].$$

The estimates of μ_i (or equivalently, of β) are not consistent if the Poisson assumption is violated (even if the link function and linear predictor are correct), because the expected value depends on correct specification of $P(y_i = 0)$.

The interpretation of regression coefficients for this model is more complicated than that for nontruncated data. If the population of interest is actually the underlying Poisson random variable without truncation (if, for example, values of the target equal to zero are possible, but the given sampling scheme does not allow them to be observed), the coefficients have the usual interpretation consistent with the log-linear model [since in that case $E(y) = \exp(\mathbf{x}'\beta)$]. If, on the other hand, observed zeroes are truly impossible, β_j no longer has a simple connection to the expected multiplicative change in y. Rather, the instantaneous expected change in y given a small change in x_j, holding all else fixed, is

$$\frac{\partial E(y_i|x_i; y_i > 0)}{\partial x_{ij}} = \beta_j \mu_i \left\{\frac{1 - \exp(-\mu_i)(1 + \mu_i)}{[1 - \exp(-\mu_i)]^2}\right\}. \qquad (10.3)$$

For nontruncated data the right-hand side of (10.3) is simply $\beta_j \mu_i$ (reflecting the interpretation of β_j as a semielasticity), so the truncation adds an additional multiplicative term that is a function of the nontruncated mean and the nontruncated probability of an observed zero. The right-hand side of (10.3) divided by μ_i represents the semielasticity of the variable for this model.

If the target variable exhibits overdispersion relative to a (truncated) Poisson variable, a truncated negative binomial is a viable alternative model. The mean of this random variable is

$$E(y_i) = \frac{\mu_i}{1 - (1 + \alpha\mu_i)^{-1/\alpha}},$$

while the variance is

$$V(y_i) = \left[\frac{\mu_i}{1 - (1 + \alpha\mu_i)^{-1/\alpha}}\right]\left[1 + \alpha\mu_i - \frac{\mu_i(1 + \alpha\mu_i)^{-1/\alpha}}{1 - (1 + \alpha\mu_i)^{-1/\alpha}}\right].$$

10.7　Poisson Regression and Weighted Least Squares

We have previously discussed how nonconstant variance related to group membership in an ordinary least squares fit can be identified using Levene's test, and handled using weighted least squares with the weights for the members of each group being the inverse of the variance of the residuals for that group. Another way to refer to nonconstant variance related to group membership is to say that nonconstant variance is related to the values of a predictor variable, where that predictor variable happens to be categorical. It is also possible that the variance of the errors could be related to a (potential) predictor variable that is numerical. Generalizing the Levene's test for this situation is straightforward: just construct a regression with the absolute residuals as the response and the potential numerical variable as a predictor. Note that this also can be combined with the situation with natural subgroups by running an ANCOVA model with the absolute residuals as the response and both the grouping variable(s) and the numerical variable(s) as predictors (of course, the response variable should not be used as one of the predictors).

Constructing weights for WLS in the situation with numerical variance predictors is more complicated. What is needed is a model for what the relationship between the variances and the numerical predictor is. An exponential/linear model for this relationship is often used, whose parameters can be estimated from the data. This model has the advantage that it can only produce positive values for the variances, as is appropriate. The model for the variance of ε_i is

$$V(\varepsilon_i) \equiv \sigma_i^2 = \sigma^2 \exp\left(\sum_j \lambda_j z_{ij}\right), \qquad (10.4)$$

where z_{ij} is the value of the jth variance predictor for the ith case and σ^2 is an overall "average" variance of the errors.

Poisson regression can be used to estimate the $\boldsymbol{\lambda}$ parameters, as is proposed in Simonoff and Tsai (2002). The key is to recognize that since $\sigma_i^2 = E(\varepsilon_i^2)$, by (10.4)

$$\log E(\varepsilon_i^2) = \log \sigma^2 + \sum_j \lambda_j z_{ij} \equiv \lambda_0 + \sum_j \lambda_j z_{ij}.$$

This has the form of a log-linear model for the expected squared errors, so a Poisson regression using the squared residuals (the best estimates of the squared errors available from the data) as the response provides an estimate of $\boldsymbol{\lambda}$, and hence estimated weights for a WLS fit to the original $\mathbf{y}$. Since (10.4) is a model for variances, the weight for the ith observation is the inverse of the fitted value from a Poisson regression for that observation. The squared residuals are not integer-valued, so it is possible that the Poisson regression package will produce an error message, but it should still provide the weights.

10.7.1 EXAMPLE – INTERNATIONAL GROSSES OF MOVIES (CONTINUED)

Recall the ANCOVA model fitting for logged international grosses of movies discussed in Section 7.3. The chosen model in that analysis was based on logged domestic grosses and MPAA rating, with different slopes and intercepts for the different rating levels. As was noted on page 145, however, there is nonconstant variance related to logged domestic grosses apparent in the residuals, with movies with lower domestic revenues having higher variability in international revenues. This suggests fitting a weighted least squares model to these data, with weights based on logged domestic grosses. A Poisson regression fit to the squared standardized residuals is

$$\widehat{(e_i^*)^2} = 2.476 - 1.606 \times \texttt{Log.domestic.gross}.$$

A WLS fit (not given) finds "Avatar" to be a leverage point due to its very high domestic gross, which corresponds to low variance and therefore high weight. Omitting "Avatar" gives the following fit for the WLS model with different slopes for different rating classes:

```
Response: Log.international.gross
                              Sum Sq  Df  F value    Pr(>F)
(Intercept)                   0.0991   1   0.3977    0.5295
Log.domestic.gross            5.4982   1  22.0652  7.184e-06 ***
Rating                        1.3839   3   1.8513    0.1417
Log.domestic.gross:Rating     1.0044   3   1.3436    0.2636
Residuals                    29.4034 118
---
Signif.
    codes:  0 '***' 0.001 '**' 0.01 '*' 0.05 '.' 0.1 ' ' 1

Residual standard error: 0.4992 on 118 degrees of freedom
Multiple R-squared: 0.6625,      Adjusted R-squared: 0.6425
F-statistic:  33.1 on 7 and 118 DF,  p-value: < 2.2e-16
```

The interaction term is not statistically significant in the WLS fitting. Omitting it and fitting a constant shift model yields an insignificant F-test for MPAA rating, and the best model (in terms of minimum AIC_c, for example) is the simple linear regression on logged domestic grosses only:

```
Coefficients:
                   Estimate Std. Error t value Pr(>|t|)
(Intercept)        -0.77623    0.18003  -4.312 3.27e-05 ***
Log.domestic.gross  1.34107    0.09107  14.726  < 2e-16 ***
---
Signif.
    codes:  0 '***' 0.001 '**' 0.01 '*' 0.05 '.' 0.1 ' ' 1

Residual standard error: 0.5056 on 124 degrees of freedom
Multiple R-squared: 0.6362,      Adjusted R-squared: 0.6333
F-statistic: 216.8 on 1 and 124 DF,  p-value: < 2.2e-16
```

Residual plots (Figure 10.7) show that the heteroscedasticity in the OLS fit has been addressed, and the model fits reasonably well.

In addition to resulting in different models with different implications (a single fitted line versus four different lines), WLS and OLS can produce very

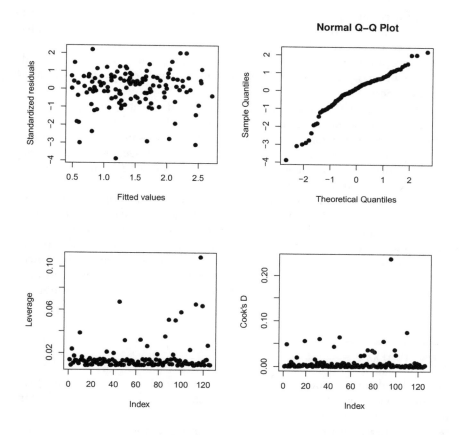

FIGURE 10.7 Diagnostic plots for WLS model for 2009 international grosses data.

different prediction intervals for some movies. For example, the 2010 movie "From Paris With Love" had a low domestic gross (and hence high estimated variability), which is reflected in the difference between the OLS prediction interval for international gross of $(0.73, 67.87)$ (in millions of dollars) and the WLS interval $(0.86, 166.41)$. By contrast, "Shutter Island" has an OLS interval of $(15.15, 1569.77)$ and a WLS interval of $(25.68, 489.68)$, reflecting its high domestic gross and implied lower variability of international grosses.

10.8 Summary

In this chapter we have used the generalized linear model framework to extend Gaussian-based least squares regression and binomial- and multinomial-based logistic regression to count data regression models. The central distribution for count regression is the Poisson distribution, with generalizations such as the negative binomial, zero-inflated, and zero-truncated distributions available to handle violations of the Poisson structure. Further discussion of generalized linear models can be found in McCullagh and Nelder (1989) and Myers et al. (2002), and more details on count data modeling can be found in Simonoff (2003).

Poisson regression models also can be fit based on categorical predictors. In the situation where all of the predictors are categorical the data take the form of a contingency table (a table of counts), and Poisson- and multinomial-based log-linear models have long been the basis of contingency table analyses. Extensive discussion of the analysis of such tables can be found in Agresti (2007), Agresti (2010), and Simonoff (2003).

The situation when the expected counts are small is termed sparse categorical data. In this situation the theory underlying standard methods does not apply, and those methods can provide misleading results. There is an extensive literature for this situation. One approach that is particularly appropriate for data with ordered categories is to use smoothing methods, which borrow information from nearby categories to improve estimation in any given category. Further discussion can be found in Chapter 6 of Simonoff (1996).

KEY TERMS

Generalized linear model: A generalization of ordinary (least squares) regression that allows for a non-Gaussian response variable and a nonlinear link between the weighted sum of the predictor variables and the expected response. Least squares, logistic, and Poisson regression are all examples of generalized linear models.

Negative binomial random variable: A discrete random variable that provides an alternative to the Poisson random variable for count regression models. One mechanism generating a negative binomial random variable is as a Poisson random variable with a random, rather than fixed, mean. It is useful for the modeling of overdispersion, as its variance is larger than its mean.

Offset: A variable included in the mean function of a generalized linear model with a specified coefficient. When in a count regression model the offset is the log of a standardizing variable such as population, and the coefficient is set to 1, this results in a model for expected rates rather than expected counts.

Overdispersion: The situation where the observed variance of a random variable is larger than would be expected given its mean. The most common example of this is for Poisson fitting of count data, since in that case the variance is assumed to equal the mean.

Poisson random variable: The most commonly-used discrete random variable used to model count data. Standard assumptions generate the Poisson random variable as appropriate for modeling the number of events that occur in time or space.

Quasi-likelihood estimation: An estimation method based on the principle of assuming a mean and variance structure of a response variable without specifying a specific probability distribution. This sometimes, but not always, leads to an estimation scheme identical to likelihood estimation for some distribution.

Zero-inflated count regression model: A generalization of count regression models such as those based on the Poisson and negative binomial distributions that allows for more zero counts than would be expected.

Zero-truncated count regression model: A generalization of count regression models such as those based on the Poisson and negative binomial distributions that does not allow for any zero counts by truncating the distribution above zero.

Nonlinear Regression

11.1 Introduction

In all of the models discussed in the previous chapters the model parameters entered linearly into the model. In those situations when it was believed that this was inappropriate we considered the use of a transformation to linearize the model. There are occasions, however, when that cannot (or should not) be done. For example, consider pharmacokinetic models, which are designed to model the course of substances administered to a living organism, such as drugs, nutrients, and toxins. A simple model for the concentration (C) in the blood of a substance as a function of time is a one-compartment model, which implies exponential decay,

$$E[C(t)] = \theta_1 e^{\theta_2 t}. \tag{11.1}$$

This model is consistent with assuming that a substance in the blood is in rapid equilibrium with the substance in other tissues, with the rate of change of concentration being directly proportional to the concentration remaining. That is, once introduced the substance mixes instantaneously in the blood and distributes throughout the body rapidly. This is simply the semilog model

(4.5), and is of course linearizable, since

$$\log\{E[C(t)]\} = \log\theta_1 + \theta_2 t.$$

Many substances, however, do not follow this pattern. Instead, the course of the substance can be thought of as being consistent with the body being made up of two compartments: the vascular system, including the blood, liver, and kidneys, where it is distributed quickly, and poorly perfused tissues, such as muscle, lean tissue, and fat, where the substance is eliminated more slowly. Such a process is consistent with a two-compartment model,

$$E[C(t)] = \theta_1 e^{\theta_2 t} + \theta_3 e^{\theta_4 t}. \tag{11.2}$$

This model is not linearizable by a transformation, so **nonlinear regression** methods are required to fit it.

Another example is the logistic model for population growth, sometimes referred to as the Verhulst-Pearl model, which is based on a differential equation. Let $P(t)$ be the population at time t, P_0 the population at the base period, M the limiting size of the population (termed the carrying capacity), and k the growth rate. Ignoring random error for the moment, under the assumption that the rate of population change is proportional to both the existing population and the amount of available resources (the room to grow), the population satisfies the differential equation

$$\frac{dP}{dt} = kP\left(1 - \frac{P}{M}\right).$$

The solution to this differential equation is

$$P(t) = \frac{MP_0 \exp(kt)}{M + P_0[\exp(kt) - 1]},$$

which cannot be linearized by a simple transformation. There are many processes in the physical sciences that arise from nonlinear differential equations, and in general nonlinear regression methods are required to fit them.

In this chapter we discuss the nonlinear regression model. Even though the model is based on least squares, the nonlinearity of the model means that the derivation and properties of inferential methods and techniques are similar to those of Chapters 8 through 10.

11.2   Concepts and Background Material

The nonlinear regression model satisfies

$$y_i = f(\mathbf{x}_i, \boldsymbol{\theta}) + \varepsilon_i. \tag{11.3}$$

The errors ε are taken to be independent and identically normally distributed, which justifies the use of least squares. The linear model (1.1) is a special case of this model, but (11.3) is obviously far more general. Note that $\mathbf{x}$ is a $p \times 1$ vector, while $\boldsymbol{\theta}$ is a $q \times 1$ vector, and p need not equal q (for example, in the two-compartment pharmacokinetic model (11.2), $p = 1$ and $q = 4$).

There are four key distinctions between nonlinear regression and the linear regression models discussed in earlier chapters, two of which can be viewed as advantages and two of which can be viewed as disadvantages.

1. Nonlinear regression is more flexible than linear regression, in that the function f need not be linear or linearizable. If the relationship between the expected response and the predictor(s) is not linear or linearizable, nonlinear regression provides a tool for fitting that relationship to the data. The only formal requirement on f is that it be differentiable with respect to the elements of $\boldsymbol{\theta}$, which implies the existence of the least squares estimates.

2. Nonlinear regression can be more appropriate than the use of transformations and linear regression in situations where f is linearizable. The reason for this is the additive form of the nonlinear model (11.3). Consider again the one-compartment semilog pharmacokinetic relationship (11.1). Fitting this relationship via the nonlinear regression model

$$C_i = \theta_1 e^{\theta_2 t_i} + \varepsilon_i$$

is consistent with a relationship exhibiting constant variance. In contrast, the semilog regression model based logging the concentration variable,

$$\log(C_i) = \beta_1 + \beta_2 t_i + \varepsilon_i$$

is equivalent to a relationship for C_i that exhibits nonconstant variance, since it implies multiplicative rather than additive errors:

$$C_i = e^{\beta_1} \times e^{\theta_2 t_i} \times e^{\varepsilon_i}$$

(this was noted earlier in Section 4.1). If the actual relationship does not exhibit heteroscedasticity, it is quite possible that fitting the model as a linear model for the logged concentration will result in a poor fit, since taking logs will induce heteroscedasticity that was not originally there. Further, the additive form (11.3) allows for the possibility of negative values of y even if f only takes on positive values; in contrast the logs of nonpositive values are undefined, so observations with $y \leq 0$ would simply be dropped from a linear regression fit for $\log y$.

3. Nonlinear regression requires knowledge of f before fitting proceeds, which implies a thorough understanding of the underlying process being examined. Linear regression models are often viewed as exploratory, being appropriate when a relationship between the response and the predictor(s) is suspected but not precisely specified; the use of residual plots

to suggest potential transformations is consistent with that exploratory
nature. Nonlinear regression, on the other hand, requires precise specification of the relationship between the response and the predictor(s),
which might be difficult or impossible.

4. If f is misspecified the fit of the regression can be extremely poor. Indeed,
depending on the form of f, it is possible that the nonlinear regression
can fit worse than no regression at all (that is, fit worse than using $\overline{Y}$ to
estimate all values of y). The price that is paid for the flexibility of f
noted above is a lack of robustness to a poor choice of f.

11.3   Methodology

Nonlinear regression methodology is based on the same principles of maximum likelihood described in Section 8.3.1. If the errors ε are Gaussian, maximum likelihood corresponds to least squares, and estimation of θ proceeds
using **nonlinear least squares**.

11.3.1 NONLINEAR LEAST SQUARES ESTIMATION

As was true for linear models, nonlinear least squares is based on minimizing
the sum of squares of the residuals,

$$S = \sum_{i=1}^{n} [y_i - f(\mathbf{x}_i, \boldsymbol{\theta})]^2.$$

The minimizer of S, $\hat{\boldsymbol{\theta}}$, satisfies the set of q equations setting the partial derivatives with respect to each entry of $\boldsymbol{\theta}$ equal to zero, but unlike for linear models
it is not straightforward to find that value. Estimation and inference can proceed through the use of a (first-order) **Taylor series approximation**, wherein
the function f is approximated at a particular value $\boldsymbol{\theta}^0$ using a linear function,

$$f(\mathbf{x}_i, \boldsymbol{\theta}) \approx f(\mathbf{x}_i, \boldsymbol{\theta}^0) + \sum_{j=1}^{q} \frac{\partial f(\mathbf{x}_i, \boldsymbol{\theta}^0)}{\partial \theta_j^0}(\theta_j - \theta_j^0). \qquad (11.4)$$

Using this approximation for f in S turns the minimization problem
into an (approximately) linear one, with the partial derivatives $\partial f(\mathbf{x}_i, \boldsymbol{\theta}^0)/\partial \theta_j^0$
taking the place of the columns of X in linear regression. This is an iterative
process, in which a current estimate of $\boldsymbol{\theta}$ is used as $\boldsymbol{\theta}^0$ in (11.4) and the residual
sum of squares S based on this linear approximation is minimized, with the
minimizer becoming the new estimate of $\boldsymbol{\theta}$. This process is continued until
a stopping criterion is met, such as a small change in S or a small change
in the estimated $\boldsymbol{\theta}$. There is no guarantee that this procedure will converge
to the global minimizer of S; indeed, not only might it converge to a local

minimizer, it can even converge to a local maximizer. A good starting value of θ increases the chance of finding the true minimizer, and will also reduce the number of iterations needed for convergence, but still provides no guarantees. For this reason it is a good idea to try different sets of starting values to see if the same solution is obtained.

Poor starting values far from the true solution are more likely to be problematic, so putting some thought into which values to try is worthwhile. For example, if a model is linearizable, the corresponding values based on a linear fit to transformed variables is a very reasonable candidate. Sometimes a special case of a model (taking particular values of θ_j equal to special values such as 0 or 1, for example) is linearizable, and estimates based on the linear fit to that special case can be worth using as starting values. It is sometimes the case that parameters have physical interpretations that can be exploited to choose starting values. For example, in the Michaelis-Menten model discussed in Section 11.4, one of the parameters represents a theoretical maximum limit for the response, so the observed maximum is a reasonable starting value to use for that parameter.

11.3.2 INFERENCE FOR NONLINEAR REGRESSION MODELS

The Taylor series approximation used when estimating θ is also the basis of large-sample inference for nonlinear regression models. Hypothesis tests and confidence intervals are based on the linear approximation, with the partial derivative evaluated at the least squares estimate $\partial f(\mathbf{x}_i, \hat{\theta})/\partial \hat{\theta}_j$ taking the place of the jth predictor column. Since the linear approximation is only valid for large samples, approximate (Wald) tests and intervals for entries of θ should perhaps be based on a normal distribution rather than a t-distribution, although many packages use the t-distribution. Nested models (where one is a special case of the other) can be compared using a partial F-test, which approximately follows an F-distribution. Diagnostics such as standardized residuals, leverage values, Cook's distances, and variance inflation factors can also be based on the Taylor series linear approximation, although this is not always available in statistical packages that provide nonlinear regression fitting. Note that this asymptotic approach is similar in spirit to the standard inferential methods for generalized linear models discussed in the previous three chapters.

On the other hand, some concepts from linear regression do not translate in a straightforward way to nonlinear regression. Since variables do not have a one-to-one correspondence with parameters, they often cannot be omitted without fundamentally changing the form of f, meaning that variable selection is not necessarily meaningful. Collinearity is as serious a problem in nonlinear regression as it is in linear regression, resulting in instabilities in estimated coefficients. Since it corresponds to high correlations among the vectors of partial derivatives, however, it is not possible to simply omit predictors to reduce collinearity. Rather, it is necessary to completely reformulate the model. Simonoff and Tsai (1989) describe how observed collinearities can be used to suggest an underlying partial differential equation that can be solved to yield a reformulation that does not exhibit collinearity, while still providing a good fit to the data. An example is provided in Niedzwiecki and Simonoff (1990), who showed that the two-compartment model (11.2) under collinearity is well-fit by the one-compartment model (11.1).

11.4   Example — Michaelis-Menten Enzyme Kinetics

The Michaelis-Menten model is a model for enzyme kinetics that relates the rate of an enzymatic reaction (the "velocity" V) to the concentration x of the substrate on which the enzyme acts. The model takes the form

$$V(x, \boldsymbol{\theta}) = \frac{\theta_1 x}{\theta_2 + x}$$

(ignoring the error term). For this model θ_1 represents the asymptotic maximum reaction rate as the substrate concentration $x \to \infty$, while θ_2 is the so-called Michaelis constant, which is an inverse measure of how quickly the rate approaches θ_1 (a smaller value of θ_2 implies that the rate is approached more quickly, indicating a higher affinity of the substrate for the enzyme).

Figure 11.1 gives scatter plots and model fits based on data from Bates and Watts (1988). The data come from an experiment relating the velocity of reaction (in counts per minute^2 of radioactive product from the reaction) to substrate concentration (in parts per million) for an enzyme treated with the antibiotic puromycin. In order to fit the Michaelis-Menten model to data, first starting values for the algorithm must be found. The model is linearizable, since

$$\frac{1}{V} = \frac{\theta_2 + x}{\theta_1 x} = \frac{1}{\theta_1} + \left(\frac{\theta_2}{\theta_1}\right)\left(\frac{1}{x}\right) \equiv \beta_1 + \beta_2 \frac{1}{x}. \qquad (11.5)$$

This means that a linear regression of $1/V$ on $1/x$ gives (initial) estimates for $\boldsymbol{\theta}$, since $\theta_1 = 1/\beta_1$ and $\theta_2 = \beta_2/\beta_1$.

Figure 11.1(a) gives a scatter plot of $1/V$ versus $1/x$, with the linear least squares line superimposed on the plot. It is clear that an OLS fit to the transformed data is not appropriate, as the transformation has induced strong heteroscedasticity. Still, the fitted least squares line ($\hat{\beta}_1 = 0.0051072$, $\hat{\beta}_2 = 0.0002472$) does not look unreasonable.

Despite this, when the linearized model is back-transformed into a fit for velocity [$\hat{\theta}_1 = 195.80$, $\hat{\theta}_2 = 0.0484$) the fit is poor for high concentrations (Figure 11.1(b)]. These values still provide good starting values for a nonlinear least squares fit, however, which converges quickly (in five iterations):

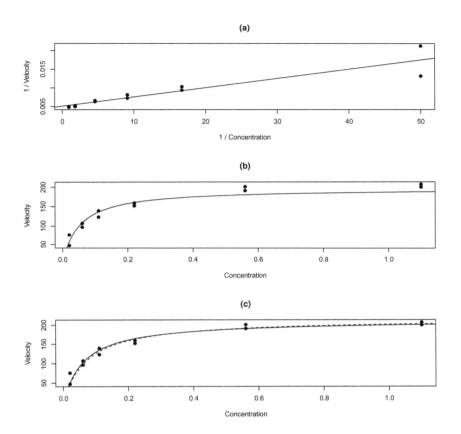

FIGURE 11.1 Scatter plots and model fits for the puromycin enzyme velocity data. (a) Scatter plot of inverse velocity versus inverse concentration, with linear least squares line superimposed. (b) Scatter plot of velocity versus concentration with fitted model based on the linear least squares model on transformed variables. (c) Scatter plot of velocity versus concentration with fitted models based on the nonlinear least squares fit using all of the data (solid line) and omitting an outlier (dashed line).

```
Parameters:
        Estimate Std. Error t value Pr(>|t|)
theta1 2.127e+02  6.947e+00  30.615 3.24e-11 ***
theta2 6.412e-02  8.281e-03   7.743 1.57e-05 ***
---

Signif. codes:
  0 '***' 0.001 '**' 0.01 '*' 0.05 '.' 0.1 ' ' 1

Residual standard error: 10.93 on 10 degrees of freedom
```

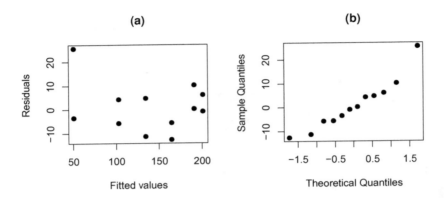

FIGURE 11.2 Residual plots for the puromycin enzyme velocity data. (a) Plot of residuals versus fitted values. (b) Normal plot of the residuals.

The estimated coefficients [$\hat{\theta}_1 = 212.68$, $\hat{\theta}_2 = 0.0641$) provide a much better fit to the data [Figure 11.1(c)], particularly in estimation of the asymptotic maximum reaction rate θ_1. Using other starting values yields virtually identical estimates. There is some evidence of an outlier in the data, corresponding to a point with unusually high velocity for a concentration of 0.02 parts per million (Figure 11.2), but omitting this observation does not change the fitted model appreciably [it is given by the dashed line in Figure 11.1(c)].

These data are part of a larger experiment in which the relationship between velocity of reaction and concentration was examined when the enzyme is not treated with puromycin. The observations corresponding to untreated enzyme are given by ×s in Figure 11.3. It was hypothesized that treatment by puromycin would affect the maximum velocity of reaction (θ_1) but not the affinity (θ_2) of the enzyme, and the effect on maximum velocity is apparent from the plot.

The full data set corresponds to a situation with two natural subgroups in the data, and is thus amenable to an analysis based on incorporating an indicator variable into the model, as was done in Section 2.4. Indeed, the linearized form of the Michaelis-Menten model (11.5) shows that a "pooled/constant shift/full" model approach is meaningful here, with the maximum velocity θ_1 taking the role of the intercept and the affinity θ_2 taking the role of the slope. If `Treated` is an indicator variable taking on the value 1 for an experiment where the enzyme was treated and 0 otherwise, the "constant shift" model adds a parameter θ_3 to the model,

$$\texttt{Velocity} = \frac{(\theta_1 + \theta_3\texttt{Treated})\texttt{Concentration}}{\theta_2 + \texttt{Concentration}},$$

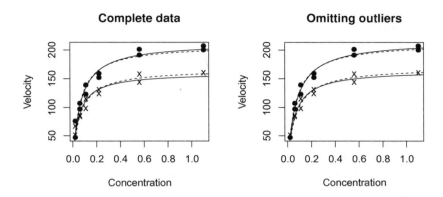

FIGURE 11.3 Scatter plots and model fits for the puromycin enzyme velocity data. Observations where the enzyme was treated with puromycin are represented by dots, while observations where the enzyme was untreated are represented by ×s. The left plot is based on all of the data, while the right plot is based on the data after removing two outliers. The solid lines represent a model with different values of θ for treated and untreated observations, while the dashed lines represent a model with different values of θ_1 and identical values of θ_2 for the two groups.

which allows for different maximum velocity values for the two groups. The full model adds a parameter θ_4,

$$\texttt{Velocity} = \frac{(\theta_1 + \theta_3\texttt{Treated})\texttt{Concentration}}{(\theta_2 + \theta_4\texttt{Treated}) + \texttt{Concentration}},$$

which allows for both different maximum velocity and different affinity depending on whether or not the enzyme is treated with puromycin.

The output below summarizes fits of the three models:

```
Parameters:
        Estimate Std. Error t value Pr(>|t|)
theta1 190.80632    8.76459  21.770 6.84e-16 ***
theta2   0.06039    0.01077   5.608 1.45e-05 ***
---
Signif. codes:
  0 '***' 0.001 '**' 0.01 '*' 0.05 '.' 0.1 ' ' 1

Residual standard error: 18.61 on 21 degrees of freedom


-----------------


Parameters:
        Estimate Std. Error t value Pr(>|t|)
theta1 166.60406    5.80743  28.688  < 2e-16 ***
theta2   0.05797    0.00591   9.809 4.37e-09 ***
```

```
theta3  42.02595     6.27214     6.700 1.61e-06 ***
---

Signif. codes:
  0 '***' 0.001 '**' 0.01 '*' 0.05 '.' 0.1 ' ' 1

Residual standard error: 10.59 on 20 degrees of freedom


------------------


Parameters:
        Estimate Std. Error t value Pr(>|t|)
theta1 1.603e+02  6.896e+00  23.242 2.04e-15 ***
theta2 4.771e-02  8.281e-03   5.761 1.50e-05 ***
theta3 5.240e+01  9.551e+00   5.487 2.71e-05 ***
theta4 1.641e-02  1.143e-02   1.436    0.167
---
Signif. codes:
  0 '***' 0.001 '**' 0.01 '*' 0.05 '.' 0.1 ' ' 1

Residual standard error: 10.4 on 19 degrees of freedom
```

The Wald (t) tests indicate that the "constant shift" model is a significant improvement over the pooled model (the test of $\theta_3 = 0$ is $t = 6.7$), but the full model is not a significant improvement over the "constant shift" model (the test of $\theta_4 = 0$ is $t = 1.4$). Just as was true for logistic regression and other generalized linear models (page 171), Wald tests are known to be less trustworthy than likelihood ratio tests (the likelihood ratio test for nonlinear regression models is a partial F-test). In this case the distinction is unimportant, since the partial F-tests ($F = 44.9$ on $(1, 20)$ degrees of freedom, $p = 1.6 \times 10^{-6}$, and $F = 1.72$ on $(1, 19)$ degrees of freedom, $p = .206$, respectively) also support the hypothesis of a different maximum velocity (estimated to be 166.6 for untreated enzyme versus 208.6 for treated enzyme) but same affinity (estimated Michaelis constant .058) for the treated and untreated groups. The left plot of Figure 11.3 gives the estimated velocity functions for the two groups based on the "constant shift" (dashed lines) and full (solid lines) models, and it is apparent that they are similar to each other.

Residual plots (not presented here) reveal two potential outliers at a concentration of .02 parts per million (one in each group), but model fitting after omitting these observations still leads to the "constant shift" model as before (and as was originally hypothesized). The right plot of Figure 11.3 gives the estimated velocity functions for the two groups based on the "constant shift" (dashed lines) and full (solid lines) models (the "constant shift" model has an estimated maximum velocity of 170.2 for untreated enzyme versus 213.2 for treated enzyme and estimated Michaelis constant .066 for both groups). It is apparent that the estimated velocities are not very different from those based on all of the data.

11.5   Summary

We have only briefly touched on the basics of nonlinear regression fitting in this chapter. Bates and Watts (1988) and Seber and Wild (1989) provide much more thorough discussion of such models, including more details on estimation. They also discuss how differential geometry can be used to construct curvature measures (intrinsic curvature and parameter-effects curvature, respectively) that quantify the extent to which the linear Taylor series approximation fails for a given data set and given model parameterization.

As was noted earlier, nonlinear regression modeling requires the strong assumption that a known nonlinear relationship is appropriate. In the situation where it is suspected that a relationship might be nonlinear, but it is not known what that relationship might be, a different approach is necessary. If it is assumed that the true underlying relationship is smooth, smoothing methods are an appropriate tool. These methods estimate the regression relationship locally, estimating $f(x)$ at the value x_0 using data from a local neighborhood around x_0. See Simonoff (1996) for further discussion of these methods.

KEY TERMS

Nonlinear least squares: A method for estimating the parameters of a nonlinear regression model. It is appropriate when the additive error term is (roughly) normally distributed with constant variance, and requires an iterative procedure to find the solution.

Nonlinear regression model: A model for the relationship between a response and predictor(s) in which at least one parameter does not enter linearly into the model.

Taylor series approximation: A method of approximating a nonlinear function with a polynomial. The first-order (linear) Taylor series approximation is the basis of standard inferential tools when fitting nonlinear regression models.

BIBLIOGRAPHY

Abzug, R., Simonoff, J.S., and Ahlstrom, D. (2000), "Nonprofits as Large Employers: A City-Level Geographical Inquiry," *Nonprofit and Voluntary Sector Quarterly*, 29, 455-470.

Agresti, A. (2007), *An Introduction to Categorical Data Analysis*, 2nd ed., John Wiley and Sons, New York.

Agresti, A. (2010), *Analysis of Ordinal Categorical Data*, 2nd ed., John Wiley and Sons, New York.

Akaike, H. (1973), "Information Theory and an Extension of the Maximum Likelihood Principle," in B.N. Petrov and F. Csaki (eds.) *2nd International Symposium on Information Theory*, Akademiai Kiado, Budapest, 267-281.

Appleton, D.R., French, J.M., and Vanderpump, M.P.J. (1996), "Ignoring a Covariate: An Example of Simpson's Paradox," *American Statistician*, 50, 340-341.

Atkinson, A. and Riani, M. (2000), *Robust Diagnostic Regression Analysis*, Springer-Verlag, New York.

Bates, D.M. and Watts, D.G. (1988), *Nonlinear Regression and Its Applications*, John Wiley and Sons, New York.

Belsley, D.A., Kuh, E., and Welsch, R.E. (1980), *Regression Diagnostics: Identifying Data and Sources of Collinearity*, John Wiley and Sons, New York.

Bretz, F., Hothorn, T., and Westfall, P. (2010), *Multiple Comparisons Using R*, Chapman and Hall/CRC, Boca Raton, FL.

Bühlmann, P. and van de Geer, S. (2011), *Statistics for High-Dimensional Data*, Springer-Verlag, New York.

Burgess, G.H. (2012), "ISAF 2011 Worldwide Shark Attack Summary," `http://www.flmnh.ufl.edu/fish/sharks/isaf/2011summary.html`. Accessed June 17, 2012.

Burnham, K.P. and Anderson, D.R. (2002), *Model Selection and Multimodel Inference: A Practical Information-Theoretic Approach*, 2nd ed., Springer-Verlag, New York.

Chatterjee, S. and Hadi, A.S. (1988), *Sensitivity Analysis in Linear Regression*, John Wiley and Sons, New York.

Chatterjee, S. and Hadi, A.S. (2012), *Regression Analysis By Example*, 5th ed., John Wiley and Sons, New York.

Chow, G.C. (1960), "Tests of Equality Between Sets of Coefficients in Two Linear Regressions," *Econometrica*, **28**, 591-605.

Cochrane, D. and Orcutt, G.H. (1949), "Application of Least Squares Regression to Relationships Containing Autocorrelated Error Terms," *Journal of the American Statistical Association*, **44**, 32-61.

Cook, R.D. (1977), "Detection of Influential Observations in Linear Regression," *Technometrics*, **19**, 15-18.

Cryer, J.D. and Chan, K.-S. (2008), *Time Series Analysis: With Applications in R*, Springer-Verlag, New York.

Durbin, J. and Watson, G.S. (1951), "Testing for Serial Correlation in Least Squares Regression. I.," *Biometrika*, **38**, 159-179.

Furnival, G.M. and Wilson, R.W., Jr. (1974), "Regressions by Leaps and Bounds," *Technometrics*, **16**, 499-511.

Greene, W.H. (2011), *Econometric Analysis*, 7th ed., Prentice Hall, Upper Saddle River, NJ.

Gregoriou, G.N. (2009), *Stock Market Volatility*, Chapman and Hall/CRC, Boca Raton, FL.

Hadi, A.S. and Simonoff, J.S. (1993), "Procedures for the Identification of Multiple Outliers in Linear Models," *Journal of the American Statistical Association*, **88**, 1264-1272.

Hilbe, J.M. (2009), *Logistic Regression Models*, Chapman and Hall/CRC, Boca Raton, FL.

Hosmer, D.W. and Lemeshow, S. (2000), *Applied Logistic Regression*, 2nd. ed., John Wiley and Sons, New York.

Hout, M., Mangels, L., Carlson, J., and Best, R. (2004), "Working Paper: The Effect of Electronic Voting Machines on Change in Support for Bush in the 2004 Florida Elections," http://www.verifiedvoting.org/downloads/election04_WP.pdf, accessed July 7, 2011.

Hurvich, C. M. and Tsai, C.-L. (1989), "Regression and Time Series Model Selection in Small Samples," *Biometrika*, **76**, 297-307.

Kedem, B. and Fokianos, K. (2002), *Regression Models for Time Series Analysis*, John Wiley and Sons, Hoboken, NJ.

Mallows, C.L. (1973), "Some Comments on C_P," *Technometrics*, **15**, 661-675.

Mantel, N. (1987), "Understanding Wald's Test for Exponential Families," *American Statistician*, **41**, 147-148.

McCullagh, P. and Nelder, J.A. (1989), *Generalized Linear Models*, 2nd ed., Chapman and Hall, London.

Montgomery, D.C., Peck, E.A., and Vining, G.G. (2012), *Introduction to Linear Regression Analysis*, 5th ed., John Wiley and Sons, Hoboken, NJ.

Myers, R.H., Montgomery, D.C., and Vining, G.G. (2002), *Generalized Linear Models With Applications in Engineering and the Sciences*, John Wiley and Sons, New York.

Niedzwiecki, D. and Simonoff, J.S. (1990), "Estimation and Inference in Pharmacokinetic Models: The Effectiveness of Model Reformulation and Resampling Methods for Functions of Parameters," *Journal of Pharmacokinetics and Biopharmaceutics*, 18, 361-377.

Pigeon, J.G. and Heyse, J.F. (1999), "An Improved Goodness of Fit Statistic for Probability Prediction Models," *Biometrical Journal*, 41, 71-82.

Prais, S.J. and Winsten, C.B. (1954), "Trend Estimators and Serial Correlation," *Cowles Commission Discussion Paper Statistics #382*.

R Development Core Team (2011), *R: A Language and Environment for Statistical Computing*, R Foundation for Statistical Computing, Vienna, Austria (http://www.R-project.org/).

Seber, G.A.F. and Wild, C.J. (1989), *Nonlinear Regression*, John Wiley and Sons, New York.

Sen, A. and Srivastava, M. (1990), *Regression Analysis: Theory, Methods, and Applications*, Springer-Verlag, New York.

Simonoff, J.S. (1996), *Smoothing Methods in Statistics*, Springer-Verlag, New York.

Simonoff, J.S. (1998a), "Logistic Regression, Categorical Predictors and Goodness-of-Fit: It Depends on Who You Ask," *American Statistician*, 52, 10-14.

Simonoff, J.S. (1998b), "Three Sides of Smoothing: Categorical Data Smoothing, Nonparametric Regression, and Density Estimation," *International Statistical Review*, 66, 137-156.

Simonoff, J.S. (2003), *Analyzing Categorical Data*, Springer-Verlag, New York.

Simonoff, J.S. and Tsai, C.-L. (1989), "The Use of Guided Reformulations When Collinearities are Present in Non-linear Regression," *Applied Statistics*, 38, 115-126.

Simonoff, J.S. and Tsai, C.-L. (2002), "Score Tests for the Single Index Model," *Technometrics*, 44, 142-151.

Strimmer, K. (2008), "A Unified Approach to False Discovery Rate Estimation," *Bioinformatics*, 9, 303.

Theus, M. and Urbanek, S. (2009), *Interactive Graphics for Data Analysis: Principles and Examples*, CRC Press, Boca Raton, FL.

Train, K.E. (2009), *Discrete Choice Methods with Simulation*, Cambridge University Press, Cambridge.

U.S. Census Bureau (2011), "The X-12-ARIMA Seasonal Adjustment Program," http://www.census.gov/srd/www/x12a, accessed August 30, 2011.

Wabe, J.S. (1971), "A Study of House Prices as a Means of Establishing the Value of Journey Time, the Rate of Time Preference and the Valuation of Some Aspects of Environment in the London Metropolitan Region," *Applied Economics*, 3, 247-255.

Weisberg, S. (1980), *Applied Linear Regression*, John Wiley and Sons, New York.

Ye, J. (1998), "On Measuring and Correcting the Effects of Data Mining and Model Selection," *Journal of the American Statistical Association*, 93, 120-131.

INDEX

CPSIA information can be obtained
at www.ICGtesting.com
Printed in the USA
BVHW04*1036110818
524187BV00001B/2/P